THE RACIAL POLICIES OF AMERICAN INDUSTRY

REPORT NO. 27

THE NEGRO IN THE ELECTRICAL MANUFACTURING INDUSTRY

by

THEODORE V. PURCELL *and* DANIEL P. MULVEY
Cambridge Center for Social Studies

Published by

INDUSTRIAL RESEARCH UNIT, DEPARTMENT OF INDUSTRY
Wharton School of Finance and Commerce
University of Pennsylvania

Distributed by

University of Pennsylvania Press
Philadelphia, Pennsylvania 19104

Library of Congress Catalog Card Number 71-157695
Manufactured in the United States of America
ISBN: 0-8122-9071-2

FOREWORD

In September 1966, the Ford Foundation announced a major grant to the Industrial Research Unit of the Wharton School to fund studies of the Racial Policies of American Industry. The purpose of the research effort, now in its fourth year, is to determine why some industries are more hospitable to the employment of Negroes than are others and why some companies within the same industry have vastly different racial employment policies, and to propose appropriate policy.

The studies have proceeded on an industry by industry basis, under the direction of the undersigned, with Dr. Richard L. Rowan, Associate Professor of Industry, as Associate Director. This study of the electrical manufacturing industry is the twenty-seventh in a series of reports dealing with specific industries. A list of industry studies already published or in process is on the back cover. In addition to the individual industry reports, major studies combining and comparing the findings of the various industry reports have been undertaken. Four of these volumes—*Negro Employment in Basic Industry, Negro Employment in Finance, Negro Employment in Public Utilities,* and *Negro Employment in Southern Industry*—have recently been published. *Negro Employment in Land and Air Transport* and *Negro Employment in Retail Trade* will be published later in 1971. This study, *The Negro in the Electrical Manufacturing Industry,* is scheduled to be included in the volume, *Negro Employment in Selected Manufacturing Industries,* which will be published at a later date.

This study, like most of our others in this series, was financed by the Ford Foundation, but by a separate grant. It is part of a more intensive and personalized study being conducted in the electrical manufacturing industry by Father Theodore V. Purcell and staff at the Cambridge Center for Social Studies, along the lines of his interesting and informative works in the meat industry (*The Worker Speaks His Mind,* 1953, and *Blue Collar Man,* 1960, both Harvard University Press). A former colleague at the Cambridge Center, Father Daniel P. Mulvey, is his co-author.

The authors wish to thank numerous management and union officials in the electrical manufacturing industry and staffs of the U.S. Office of Federal Contract Compliance and Equal Employment Opportunity Commission, for their generous assistance. Miss Irene Wylie contributed much toward the research, analysis, and writing. Miss Catherine Musham checked many details, Miss Donna Woodcock typed the original manuscript, and Mrs. Veronica Kent the final draft. Considerable data were computerized and put in tabular form by Messrs. Lester Rubin and Robert C. Gulledge and Miss Elsa Klemp of the Industrial Research Unit staff. The final manuscript was edited by Mrs. Marie R. Keeney, who also prepared the index.

As in most previous reports, the data cited as "in the authors' possession" have been carefully authenticated and are on file in the Industrial Research Unit library.

HERBERT R. NORTHRUP, *Director*
Industrial Research Unit
Wharton School of Finance and Commerce
University of Pennsylvania

Philadelphia

April 1971

TABLE OF CONTENTS

LIST OF TABLES

LIST OF FIGURES

CHAPTER I

Introduction

The electrical machinery industry is a newcomer to racial integration. Although the industry itself dates back to the 1870's when Edison developed his lamp and power generators, very few Negroes were employed before World War II. Even today, the industry lags behind American industry as a whole in Negro employment. Why this is so and what changes are likely in the future are examined in this study.

Historically, black employment was limited because the industry was located in the smaller northeastern cities where labor forces prior to World War II were almost entirely white. The industry's rapid growth, however, led to geographical diversification which, combined with Negro migration to the North, has lessened the restrictive effects of location. Nevertheless, locational factors continue to have unfavorable aspects in regard to Negro employment.

Expansion has meant steady increase in the demand for labor. As in many other industries, the proportion of white collar and professional jobs in the electrical industry is growing rapidly. Production methods, however, continue to require a large number of semiskilled employees, thus creating an increasing number of blue collar jobs as well.

The electrical machinery industry has recently been in a position to assume a leading role in the employment of Negroes. Despite a favorable management stance, little progress was made before the mid-sixties. The concept of nondiscrimination, even when genuinely endorsed by management, did not overcome the patterns of recruitment that tended to perpetuate a predominantly white work force.

With a new stress on affirmative action, there is now evidence that this situation has altered considerably. We shall examine the current situation after an analysis of those aspects of the industry which affect Negro employment and a description of the situation prior to 1960.

CHAPTER II

The Electrical Manufacturing Industry

The electrical manufacturing industry is most diversified. Products range from radios, televisions, and communication equipment to electronic components, industrial and transportation control devices, and household appliances such as refrigerators, ranges, and washing machines. Wiring devices, lights of all sizes, tubes, switchgears, transformers, motors, batteries, huge turbines, and tiny diodes and transistors are all included.

Defining an industry is not easy. In many cases there are shadow areas where it is difficult to distinguish between the product of one industry and another. Industries are defined in a number of ways: the iron and steel industry in terms of raw materials; chemicals by the process of production; transportation equipment by the use of the end product. Electrical manufacturing is identified by a common power source with which products of the industry are related: electrical energy.

This common power source is the basis for the Standard Industrial Classification of the U.S. Office of Statistical Standards: "establishments engaged in manufacturing machinery, apparatus, and supplies for the generation, storage, transmission, transformation, and utilization of electrical energy." [1] Establishments are individual plants and offices classified by product regardless of the primary business of the company. Electrical manufacturing is designated SIC No. 36, and within this general two digit category, numerous three and four digit subcategories are established, indicative of the diversity of products included within the industry category. When we present industry statistical data, the material will generally be drawn from the SIC No. 36 category.

Since our prime concern is corporate policy and practice affecting minority employment, we also present company data based on all the plants in a given company, even though some plants

[1] U.S. Bureau of the Budget, *Standard Industrial Classification* Manual (Washington: Government Printing Office, 1967), p. 168. We shall use the terms "electrical machinery," "electrical manufacturing," and "electrical equipment and supplies" to describe the same industry.

produce goods which do not fall within SIC 36. In using company data, we select those firms which are primarily engaged in electrical manufacturing. For example, General Electric and Westinghouse are designated "electrical manufacturers" although they both manufacture many non-electrical products such as jet engines, steam turbines, and chemicals. On the other hand, we exclude such corporate giants as General Motors, Ford, and Union Carbide, although they are important producers of some electrical products, because their primary interests are elsewhere. We also exclude companies that are primarily business machine manufacturers, such as International Business Machines and Sperry Rand, although electrical manufacturers such as Honeywell, RCA, and until recently General Electric, have divisions engaged in manufacturing computers. Finally, we exclude conglomerates such as Litton, International Telephone and Telegraph, and General Telephone and Electronics, all of which have sizeable electronic and electrical manufacturing divisions, but which we believe are not primarily electrical manufacturers. Obviously any demarcation is somewhat arbitrary and ours is hopefully no more so than others might be.

INDUSTRIAL STRUCTURE

The electrical industry is one of the largest American industries. Table 1 shows that in 1967 it handled shipments valued at over 43 billion and employed more than 1.8 million people, over 9 percent of the total employment in manufacturing.

TABLE 1. *Electrical Manufacturing Industry and All Manufacturing*
Number of Employees, Value Added, and Value of Shipments
1967

Industry	Total Employees	Value Added by Manufacture (In Thousands)	Value of Shipments (In Thousands)
All manufacturing	19,398,000	$259,301,000	$554,564,000
Electrical machinery	1,884,000	24,855,000	43,606,000
Electrical percentage	9.7	9.6	7.9

Source: U.S. Bureau of the Census, *1967 Census of Manufactures*, Preliminary Data (April 1969).

The leading position of the electrical industry in relation to other two digit classifications is apparent in Table 2. Electrical is fourth of the seven leading American industrial groups in adding value to its products. It is *second* in employment. Moreover, as we shall indicate later, the growth rate of electrical manufacturing is such that the industry is likely to move upward in the future. The only "industry" that exceeds electrical machinery in employment is "transportation equipment," a statistical amalgam of automobiles, aerospace, shipbulding, and other diverse industries which in fact have little in common.

Large Firm Dominance

The size of the electrical manufacturing industry is further exemplified by the data in Table 3. The four largest companies—General Electric, Western Electric, Westinghouse, and RCA—employed 873,619 persons in 1969—5.9 percent of the 14,813,809 employed by the 500 largest manufacturing companies; the fif-

TABLE 2. *Electrical Manufacturing and Other Industries Number of Employees and Value Added 1967*

SIC Code	Industry Group	Total Employees	Rank	Value Added (In Thousands)	Rank
37	Transportation equipment	1,890,000	1	$28,901,000	1
36	Electrical machinery	1,884,000	2	24,855,000	4
35	Machinery, except electrical	1,872,000	3	27,697,000	2
20	Food	1,654,000	4	26,352,000	3
34	Fabricated metals	1,307,000	5	17,054,000	7
33	Primary metals	1,283,000	6	20,148,000	6
28	Chemicals and allied products	854,000	7	23,440,000	5

Source: U.S. Bureau of the Census, *1967 Census of Manufactures*, Preliminary Data (April 1969).

TABLE 3. *The Fifteen Largest Electrical Manufacturing Companies, 1969 Statistics*

Company and 1969 Rank among Industrial Corporations		Headquarters	Sales (000)	Assets (000)	Net Income (000)	Invested Capital (000)	Number of Employees	Net Income as a Percent of Sales	Net Income as a Percent of Invested Capital
General Electric	(4)	New York	$8,447,965	$6,007,490	$278,015	$2,539,977	400,000	3.3	10.9
Western Electric	(11)	New York	4,883,221	3,171,867	227,025	2,006,248	203,608	4.6	11.3
Westinghouse Electric	(17)	Pittsburgh	3,509,153	2,477,611	149,888	1,389,330	142,011	4.3	10.8
RCA	(21)	New York	3,187,903	2,634,379	151,283	1,023,754	128,000	4.7	14.8
Singer	(45)	New York	1,902,144	1,438,615	77,721	680,914	133,000	4.1	11.4
Honeywell	(75)	Minneapolis	1,425,993	1,222,011	62,481	494,115	81,520	4.4	12.6
Raytheon	(86)	Lexington, Mass.	1,285,134	538,836	35,232	259,825	53,300	2.7	13.6
Whirlpool	(100)	Benton Harbor Mich.	1,153,530	573,929	45,943	250,654	30,301	4.0	18.3
Motorola	(132)	Franklin Park, Ill.	873,224	576,509	33,793	326,134	45,000	3.9	10.4
Texas Instruments	(135)	Dallas	831,822	526,758	33,511	281,548	58,974	4.0	11.9
Zenith Radio	(166)	Chicago	676,577	336,525	39,621	224,814	22,500	5.9	17.6
Emerson Electric	(177)	St. Louis	628,427	397,380	49,935	280,526	25,800	7.9	17.8
McGraw-Edison	(183)	Elgin, Ill.	609,102	360,706	32,971	280,398	25,000	5.4	11.8
North American Philips	(199)	New York	548,960	418,785	27,424	218,452	18,000	5.0	12.6
Magnavox	(204)	New York	539,843	284,824	39,107	173,347	20,100	7.2	22.6

Source: *Fortune,* Vol. LXXXI, No. 5 (May 1970), pp. 184-191.

teen largest employed 9.4 percent of the total employment of the 500.[2]

Like many basic American industries, the electrical industry is dominated by a small number of corporate giants. Although there are approximately 10,000 American companies in the electrical industry, the eight largest firms produce more than 90 percent of total sales in about a third of the electrical product groupings. In only seven groups do they produce less than 50 percent. The eight largest, for example, produce 91 percent of the household refrigerators and 98 percent of the batteries.[3]

Employment is similarly concentrated. With total employment in the industry close to two million, 90 percent of the companies hire fewer than 100 employees. Only 167 companies employ more than 1,000 people.[4] The "Big Four" with more than 800,000 employees, although not all working in the electrical industry, have a large minority of total electrical employment.

The concentration of an industry in a few key firms gives to the top managements of those firms opportunities for leadership far greater than those of the small or even the middle-sized business management. Clearly their policies directly affect greater numbers of people and are most widely noted by other businessmen.

Great company size also allows managers greater flexibility in planning. Table 3 shows the managers of the Big Four controlling more than $14 billion in assets. The managers of such resources can afford to make larger investments of time and money for long-term benefits and are therefore more frequently concerned with strategic decisions involving long-run growth, market share, and the security of the enterprise. These managers have long-term perspectives which they can apply to manpower planning and to the education and upgrading of the labor force, including Negroes. They also have more resources to commit to the effort.

The considerable size of the electrical industry means that a large number of jobs are potentially available to Negroes. Turnover alone opens tens of thousands of jobs a year. Size is also

---

[2] *Fortune*, Vol. LXXXI, No. 5 (May 1970), pp. 184-200.

[3] *Concentration Ratios in Manufacturing* (Washington: U.S. Government Printing Office, 1967).

[4] Data in authors' possession.

important because a large industry often exercises a leadership role. Eckstein and Wilson identify electrical machinery as one of the key groups in collective bargaining which tends to influence wage rates in other industries.[5] Similarly, the electrical industry's policies and practices regarding Negro employment may influence patterns developing elsewhere in the economy, although such influence has been found by studies in this series to be limited by the industrial and manpower requirements of different industries.[6]

Growth Rate

Size gives a basic static dimension to an industry, but growth rate is a fundamental dynamic measure. Rowan, in his study, *The Negro in the Steel Industry,* points up the impact of rate of growth on employment of blacks. Employment in the basic steel industry is over half a million, but basic steel has not been a growth industry for the last ten years. Rowan asks: "How has an industry that is declining in terms of employment and beset by financial, production, and marketing problems adjusted to a contemporary problem of employing and upgrading the Negro sector of the labor force?"[7] Rowan finds that despite management's desire to improve minority employment patterns, the state of the steel industry has made improvement in employment and upgrading most difficult. In effect, industry growth in sales and employment is almost always a prerequisite for solid accomplishment in placing and upgrading Negroes.[8] The growth rate of electrical manufacturing is notable and enhances equal employment opportunities.

---

[5] Otto Eckstein and Thomas A. Wilson, "The Determination of Money Wages in American Industry," *Quarterly Journal of Economics,* Vol. LXXVI (August 1962), pp. 384-385, 394.

[6] For an analysis of the industry factors affecting Negro employment, see Herbert R. Northrup, Richard L. Rowan, *et al., Negro Employment in Basic Industry,* Studies of Negro Employment, Vol. I (Philadelphia: Industrial Research Unit, Wharton School of Finance and Commerce, University of Pennsylvania, 1970), Part One, Chapter I, and Part Eight, Chapter I.

[7] *Ibid.,* Part Four, p. 234.

[8] The urban transit and shipbuilding industries have been exceptions to this rule. See Report No. 17 and No. 18 in the Racial Policies of American Industry series.

Backman, writing of electrical manufacturing in the early 1960's, reports somewhat exuberantly of the industry's prior three decades:

> Since 1923 the growth (of the electrical manufacturing industry) has continued at an almost unbelievable rate. In thirty-five years (1958), value added by manufacture of electrical machinery has risen to $9.2 billion, more than eleven times as large as the 1923 total of $800 million—a tremendous increase even though in part this rise reflected the price inflation of the period. Total employment in 1958 was more than three times as large as in 1923. As compared with an output valued at $1.25 billion in 1923, total sales were $29.9 billion in 1958. The number of establishments increased from 1,782 in 1923 to 7,066 in 1958. Electric power consumption increased from 42.3 billion kilowatt hours in 1923 to 683.2 billion kilowatt hours in 1960, a sixteen-fold increase—and this is in physical terms and hence does not reflect price inflation.
>
> Regardless of the measurement used, the "extraordinary development" of the industry has continued despite the wide cyclical swings noted. The electrical machinery industry grew so rapidly after 1923 that by the late 1950's its relative share had doubled as compared with both all manufacturing and the entire economy. The electrical machinery industry has been one of the great growth industries in the United States. The beneficial effects of this growth have been reflected in more job opportunities, increasing productivity in our economy, a lightening of the worker's task and of the drudgery of the housewife, and better living for all of us as consumers.[9]

In the years between 1961 and 1965, the electrical machinery industry continued to achieve increases in output well above the average, along with non-electrical machinery, instruments, primary metals, fabricated metals, lumber and wood products, transportation equipment and ordnance, rubber and plastics, and chemicals.[10] The *Fortune* 500 in 1969 shows a 9 percent increase in sales and a 18.2 percent increase in profits in the appliances and electronic industry.[11]

This healthy growth is quite likely to continue. Present electrical energy capacity may well be doubled by 1980.[12] An increase

---

9 Jules Backman, *The Economics of the Electrical Machinery Industry* (New York: New York University Press, 1962), p. 57.

10 "The Production Expansion in Perspective," *Survey of Current Business*, Vol. 46, No. 1 (January 1966), p. 14.

11 *Fortune*, Vol. LXXXI, No. 5 (May 1970), pp. 200-201.

12 Edison Electric Institute, *Historical Statistics of the Electrical Utility Industry* (New York: The Institute, 1962), Section 1, p. 1. See also Edison Electric Institute, *Statistical Year Book of the Electric Utility Industry, 1969.*

in production of electrical energy will generate demand for capital goods such as electrical generators and transmission and distribution equipment. Production of consumer goods, such as household appliances, is more subject to cyclical fluctuation, but growth in population, increases in per capita income, plus rising consumer aspirations are likely to maintain or to raise the demand for consumer goods.

Further, the technology connected with electrical power has been one of the most rapidly advancing sectors of the engineering field. The industry has been intimately concerned with four of the basic technological innovations of the century: electricity as a source of productive power, radio-television communications, electronic computers and servo-mechanisms, and nuclear power generation. The industrial products and consumer goods applications forthcoming from these innovations have not yet been exhausted if, indeed, their major potential has even been tapped. The evolution of new uses and new products connected with these innovations will make for growth in the electrical manufacturing industry both in sales and employment, and therefore opportunities for increased Negro employment.

There are three cautions, however: the close association of the electrical manufacturing industry with government and, in particular, with defense production; the rising tide of imports; and the growing conflict over locations for power generating stations. In some sectors of the industry, almost one-half the total shipments are for military order or for space or areospace usage. Several of the largest electrical machinery companies are large defense contractors and for some of these firms, the value of defense contracts alone amounts to 20 percent of their total sales. For Raytheon, defense amounted to 39 percent in 1969.[13] Since 1969, our defense commitments have drastically declined. This has already significantly reduced production in certain areas of electrical manufacturing.

The import problem is already severe. Few radios are now manufactured in this country, and Japan's rising electronics technology threatens American jobs in appliances, wiring devices, television and electronic circuitry, among other branches of the industry. Finally, the growing conflict between conservationists

[13] William Simon Rubseyser, "Where the Military Contracts Go," *Fortune*, Vol. LXXX, (August 1, 1969), p. 74; and Raytheon annual report, 1969.

and public utilities in regard to the location of power generating plants could slow expansion of electrical manufacturing, retarding the demand for turbines, generators, and other apparatus.

MANPOWER

Table 4 traces the history of employment in the electrical machinery industry since 1940. During World War II, employment in the industry more than doubled; and in the postwar period, although it declined slightly, it was buoyed up almost to wartime levels by pent-up consumer demand for appliances and industrial demand for utility apparatus equipment, motors, and control mechanisms. The 1950's saw an uneven advance in jobs, but a substantial one nevertheless, amounting to nearly one-half million jobs. In the 1960's, a slightly larger total increase brought jobs in the industry to over 2 million by the end of the decade. Although employment has declined in some years, and suffers when general demand slackens, the trend is definitely upward.

As a matter of fact, the electrical manufacturing industry has accounted for a constantly growing share of both total manufacturing and durable goods employment as shown in Figure 1. Between 1947-1969, employment in all manufacturing increased from 15,545,000 to 20,121,000, a gain of 29.4 percent, and employment in durable goods manufacturing increased from 8,385,000 to 11,880,000, an increase of 41.7 percent. In electrical manufacturing, however, employment in the same period rose from 1,035,000 to 2,038,000, an increase of 96.9 percent. If past growth rates are projected forward, the industry may expect an increase of approximately 45,000 new jobs per year during the decade of the 1970's.[14] Of course, declining defense requirements and rising imports could dampen this forecast.

---

[14] U.S. Bureau of Labor Statistics, *Employment and Earnings, 1909-68*, pp. 48, 54, and 302, and *Employment and Earnings*, March 1970, Table B-2. This extrapolation is based on the post-World War II trend in the industry from 1947 to 1969. Such an average increase should not be anticipated every year. But our estimate is modest. In the decade of the sixties, annual changes of employment in the electrical industry ranged from a drop of 13,000 (1962-1963) to a rise of 250,000 (1965-1966), and the average annual increase in this decade was 64,100.

TABLE 4. *Electrical Manufacturing Industry Total, Female, and Production Worker Employment Selected Years, 1940-1969*

Year	All Employees	Female Employees: Total	Female Employees: Percent of all Employees	Production Workers: Total	Production Workers: Percent of all Employees
	(In Thousands)			(In Thousands)	
1940	494	n.a.	n.a.	375	75.9
1944	1,087	n.a.	n.a.	878	80.8
1950	991	367	37.0	770	77.7
1952	1,185	460	38.8	909	76.7
1953	1,333	528	39.6	1,029	77.2
1954	1,190	442	37.1	884	74.3
1956	1,323	496	37.5	975	73.7
1957	1,344	496	36.9	959	71.3
1958	1,249	449	35.9	857	68.6
1960	1,467	535	36.5	996	67.9
1962	1,567	589	37.6	1,051	67.1
1964	1,544	577	37.4	1,036	67.1
1966	1,909	773	40.5	1,325	69.4
1968	1,963	782	39.8	1,312	66.8
1969	2,038	820	40.2	1,357	66.6

Source: U.S. Bureau of Labor Statistics, *Employment and Earnings Statistics for the United States, 1909-68,* Bulletin 1312-6 (August, 1968), pp. 302-303; and *Employment and Earnings,* March 1969 and 1970, Tables B-2 and B-3.

FIGURE 1. *Electrical Manufacturing Industry Employment as a Percent of All Manufacturing and Durable Goods Manufacturing 1947-1969*

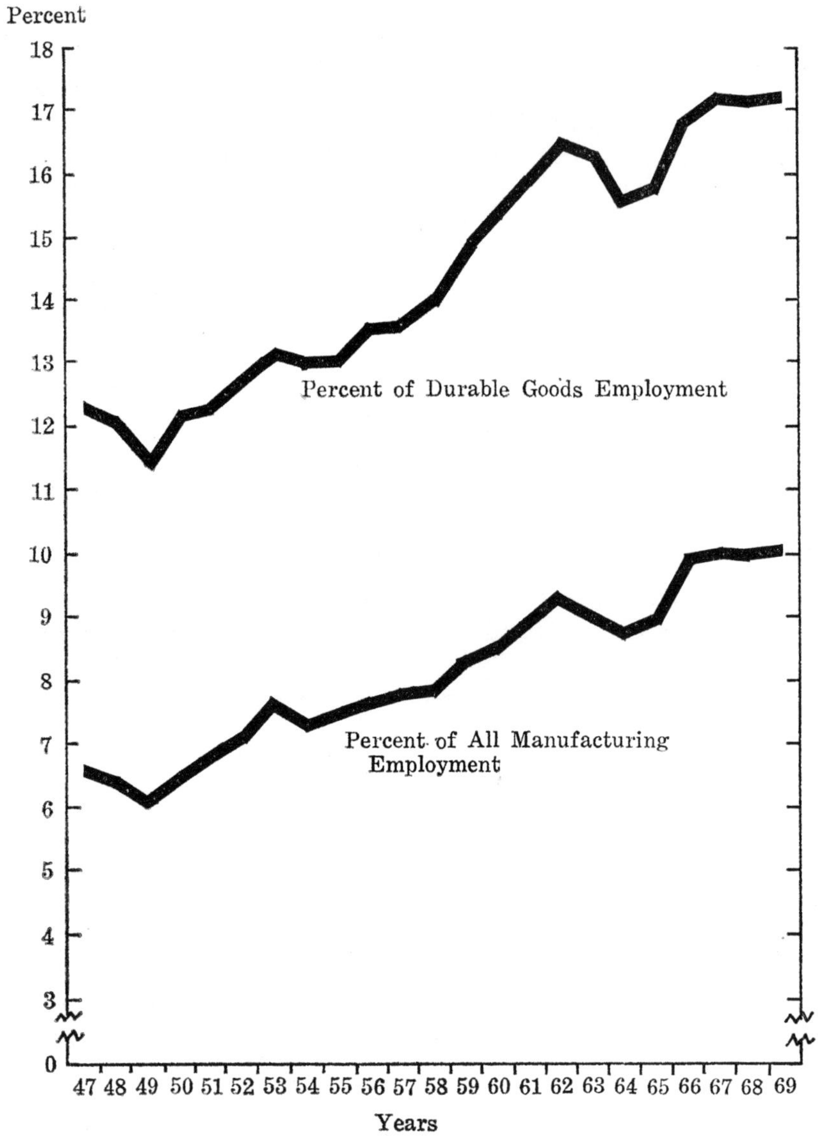

Source: U.S. Bureau of Labor Statistics, *Employment and Earnings Statistics for the United States, 1909-68*, pp. 48, 55-56, 302; and *Employment and Earnings*, March 1969 and 1970, Table B-2.

Labor Turnover

Table 5 shows the rate of accessions and separations per 100 employees in the industry as compared with all manufacturing for selected recent years. Turnover is lower in the electrical industry than in all manufacturing, indicating that electrical manufacturing jobs are considered generally desirable. There are, however, sufficient separations to provide a substantial number of openings per year. If we estimate, for example, a separation rate of 3.5 per hundred employees and a layoff rate of 1.2, which have been the averages for the industry since 1958 (Table 5), we can estimate that 2.3 new employees per hundred employed are required to maintain labor requirements in the industry. Based on 1969 employment, that would mean that normal turnover creates approximately 4,700 new jobs in the industry annually (and of course more as employment expands).

TABLE 5. *Electrical Manufacturing Industry and All Manufacturing, Turnover Data, 1958-1969*

	Electrical Manufacturing					All Manufacturing				
	Accessions		Separations			Accessions		Separations		
Year	Total	New Hires	Total	Quits	Lay-offs	Total	New Hires	Total	Quits	Lay-offs
	(Per 100 Employees)									
1958	3.3	1.5	3.5	1.0	2.1	3.6	1.7	4.1	1.1	2.6
1959	4.0	2.6	3.2	1.4	1.2	4.2	2.6	4.1	1.5	2.0
1960	3.2	2.0	3.5	1.2	1.6	3.8	2.2	4.3	1.3	2.4
1961	3.6	2.1	3.3	1.2	1.4	4.1	2.2	4.0	1.2	2.2
1962	3.6	2.4	3.3	1.4	1.1	4.1	2.5	4.1	1.4	2.0
1963	3.1	1.9	3.4	1.3	1.4	3.9	2.4	3.9	1.4	1.8
1964	3.3	2.1	3.2	1.2	1.2	4.0	2.6	3.9	1.5	1.7
1965	3.9	2.9	3.1	1.6	0.8	4.3	3.1	4.1	1.9	1.4
1966	4.7	3.8	3.8	2.3	0.5	5.0	3.8	4.6	2.6	1.2
1967	3.6	2.5	4.0	2.0	1.1	4.4	3.3	4.6	2.3	1.4
1968	3.7	2.7	3.8	2.0	0.8	4.6	3.5	4.6	2.5	1.2
1969	4.0	3.1	4.0	2.3	0.7	4.7	3.7	4.9	2.7	1.2

Source: U.S. Bureau of Labor Statistics, *Employment and Earnings Statistics for the United States, 1909-68,* Bulletin 1312-6 (August, 1968), pp. 52-54 and 304-305; and *Employment and Earnings,* April 1969 and March 1970, Table D-2.

Occupational Distribution

Despite its scientific advances and the consequent corps of professional and technical personnel employed in the industry, the electrical manufacturing industry continues to need a large percentage of production workers. Table 4 shows that in 1940, 75.9 percent of the employees in the industry were classified as production workers; in 1969 this percentage had declined to 66.6. With two-thirds of the employment in production, there remains a substantial number of hourly rated jobs in the industry.

Growth in the industry has thus had a substantial and rounded impact on employment. Unlike other growth industries with high technology, such as petroleum and chemicals, electrical manufacturing remains labor intensive—that is, the ratio of labor to capital is relatively high. The *Fortune* survey for 1969 reports $15,028 of assets committed for each employee in appliances and electronics, putting the electrical industry nineteenth of the twenty-two industries listed, far outstripped by such industries as petroleum refining with $98,403 for each employee, chemicals with $31,417 for each employee, and even farm and industrial machinery with $19,555.[15]

The reason for this is that in many sectors of electrical manufacturing, hand assembly is still a common method of manufacture. In the capital goods area, such as turbine generators and transmission equipment, production is frequently engineered and manufactured for individual specification. The production of electronic circuitry with sophisticated designs and rigid tolerances requires hand labor and human skills. Even in household appliances and radio or television assembly, where automation and printed circuitry have replaced much hand labor, inspection, hand assembly, and parts manufacturing are still necessary and provide many jobs.

A more detailed look at occupational distribution in the industry is provided by Table 6, which shows the occupational grouping of employees in 1969 based on reports to the Equal Employment Opportunity Commission. Two important facts in Table 6 stand out. On one hand, we find a high percentage (38.6) of employees in the electrical industry engaged in white collar occupations. Negro participation has always been lowest in white collar jobs, especially those in the managerial, profes-

[15] *Fortune*, Vol. LXXXI, No. 5 (May 1970), p. 203.

sional and technical categories. Great efforts will be needed to increase minority employment in such jobs.

The second fact is that the largest single occupational group in the electrical industry is semiskilled "operatives": 38.0 percent of all employees in the industry, together with laborer and service jobs, almost 50 percent of the total. Job training for these positions can be relatively brief and turnover in this category is frequently high. It is here that industry can most readily find places for new black employees. Since over one-half of these jobs are held by women, more job opportunities are open to black women than to black men, although jobs for the Negro man are a critical issue in minority employment.

Table 6 shows that 12.1 percent of the jobs in the electrical industry are in the skilled craftsmen category. Negroes have always experienced difficulty in entering these occupations, in considerable part because of union discrimination in the building and metal trades where many craftsmen learn their trades.

Unskilled labor and service occupations together account for 11.3 percent of employment. In electrical manufacturing, as in other industries, the Negro has been most heavily represented in these categories. Labor and service jobs do, however, provide opportunities for low-skilled or disadvantaged blacks, although today's emphasis is on upward movement. Yet these jobs often lead to better ones.

Looking again at the white collar occupations, we find that almost one-third of all white collar positions are clerical. Clerical training is relatively easy to supply and turnover is high. We can expect that Negroes will be employed in increasing numbers in this classification.

Technicians comprise 7.1 percent of the electrical work force. Technical requirements eliminate large numbers of less educated blacks, although many Negroes are now getting more technical training.

Professionals, 10.1 percent of total industry employment, are primarily engineers. Given the present vast shortage of Negro engineers, scientists, and mathematicians (only about 750 graduates each year), it will be decades before Negroes are proportionately represented among the professionals. Moreover, since upper level managers are frequently drawn from the professional ranks, the same time lag can be expected in the upper echelons

TABLE 6. *Electrical Manufacturing Industry*
Employment by Occupational Group
3,533 Establishments, United States, 1969

Occupational Group	All Employees		Male		Female	
	Number	Percent of Total	Number	Percent of Men	Number	Percent of Women
Officials and managers	137,731	7.9	134,087	12.3	3,644	0.6
Professionals	176,851	10.1	171,489	15.8	5,362	0.8
Technicians	124,664	7.1	113,766	10.4	10,898	1.6
Sales workers	24,438	1.4	21,724	2.0	2,714	0.4
Office and clerical	211,348	12.1	63,058	5.8	148,290	22.5
Total white collar	675,032	38.6	504,124	46.3	170,908	25.9
Craftsmen	210,937	12.1	194,336	17.9	16,601	2.5
Operatives	664,749	38.0	289,418	26.6	375,331	56.9
Laborers	169,545	9.7	77,027	7.1	92,518	14.1
Service workers	27,411	1.6	23,316	2.1	4,095	0.6
Total blue collar	1,072,642	61.4	584,097	53.7	488,545	74.1
Total	1,747,674	100.0	1,088,221	100.0	659,453	100.0

Source: Preliminary Equal Employment Opportunity Commission data, 1969.

of management. Yet the breakthrough even of a small number of Negroes to higher level administrative positions will represent an important change in traditional employment patterns.

Female Employment

We have already noted that jobs in the industry may be easier to secure for Negro females than males. Table 4 shows that about 40 percent of the total labor force is female. Table 6 shows that females are concentrated in two occupational groups—clerical and operatives. The former is, of course, typical of all industry. The latter is the result of the large number of light assembly and small apparatus wiring jobs which exist in the industry.

Electronic advances and the miniaturization of equipment since World War II have created demands for thousands of female assemblers. Expansion in this sector of the labor force has been substantial in recent years and promises to continue. Typically, such jobs require little training and are now readily open to black women.

Earnings and Working Conditions

Working conditions in the industry are varied, but are becoming increasingly more pleasant. Most older plants of the major producers have some of the drawbacks of age, but are well kept up and modernized. Newer ones of all companies, however, tend to be in pleasant surroundings, well laid out and often air conditioned because electronic components and instruments now require strict temperature and dust control in the manufacturing process. With the new plants usually go a suburban or rural location and therefore transit problems for urban dwelling Negroes. Wages are quite good, although below some other major manufacturing industries, as noted in Table 7.

Unionization

Unlike the steel, automotive, and rubber industries, the electrical industry is not organized by a single powerful and cohesive union. The International Union of Electrical, Radio, and Machine Workers (IUE) has the largest membership—315,000 or about 30 percent of the blue collar employment. The United Electrical, Radio and Machine Workers (UE) follows with 165,000 members. The International Brotherhood of Electrical Workers (IBEW), the International Association of Machinists and Aero-

TABLE 7. *Electrical Manufacturing and Other Industries Average Hourly and Average Weekly Earnings 1969*

Industry Group	Average Hourly Earnings	Average Weekly Earnings
All manufacturing	$3.19	$129.51
Durable goods	3.38	139.59
Nondurable goods	2.91	115.53
Fabricated metal products	3.33	138.53
Electrical equipment and supplies	3.09	124.84
Transportation equipment		
Motor vehicles and equipment	4.10	170.97
Aircraft and parts	3.87	161.77

Source: U.S. Bureau of Labor Statistics, *Employment and Earnings*, March 1970, Table C-2.

space Workers (IAM), and the United Automobile, Aerospace and Agricultural Implement Workers (UAW) also represent substantial numbers, while smaller groups are represented by several other unions.

The reasons for this multiplicity of union affiliation are found in history. The IBEW, the oldest union in the field, was primarily a building and metal trades craft organization, and when it did extend membership to non-craft factory workers, it offered them only "Class B" membership—one vote per local instead of one per individual in national union affairs. The claims of other craft unions in the American Federation of Labor also precluded successful organization before 1933 despite the willingness of President Gerard T. Swope and Chairman Owen D. Young, of the General Electric Company. Thus Northrup states:

In 1926, Swope actually arranged a meeting with William Green, President of the American Federation of Labor, at which Young was also present. Swope proposed that Green establish an industrial union for electrical manufacturing employees. Swope noted in this conversation that the General Electric employee representation plan could be a nucleus for such an organization, but emphasized General Electric's poor World War I experience with, and its firm opposition to dealing with a conglomeration of craft unions with their internal jealousies and jurisdictional problems. Green was, of course, helpless to act because of craft union control of AFL policies, and never even communicated with Swope after the meeting.[16]

After the inauguration of Franklin D. Roosevelt as President in 1933, unions sprang up in the electrical industry, as in other centers, and rather easily won management recognition because of the policy of General Electric, then as now the largest concern, to recognize unions chosen by their employees and not to oppose such choice. Aided by this policy, the United Electrical, Radio and Machine Workers (UE) affiliated with the newly formed Committee for Industrial Organization (CIO), organized by the late John L. Lewis, and following the passage of the National Labor Relations (Wagner) Act, won bargaining rights in most plants of the large producers, except for those of Western Electric.

The UE, however, was successfully infiltrated and controlled by adherents of the Communist Party line,[17] and was expelled from the CIO in 1949. The CIO simultaneously chartered the rival IUE, which became the dominant union in the industry. IUE, however, neither totally eliminated UE, nor garnered all its organized plants. The IBEW won several, and is now very strong in Westinghouse and is the dominant union in RCA and Western Electric. The Communications Workers of America, with a strong base in the Bell System, also represents employees in Western Electric; as noted, many other unions, such as IAM and UAW, are also now active in the industry. Perhaps more significant is the fact that nonunion plants are relatively common for manufacturing industry, and General Electric, which once welcomed unions into its plants, has since 1946 followed the vigorous collective bargaining policies initiated by its now retired vice president of employee relations, Lemuel R. Boulware.[18]

---

16 Herbert R. Northrup, *Boulwarism* (Ann Arbor: Bureau of Industrial Relations, University of Michigan, 1964), p. 13. See also David Loth, *Swope of GE* (New York: Simon and Shuster, 1950), pp. 168-172, and Frances Perkins, *The Roosevelt I Knew* (New York: The Viking Press, 1946), p. 309.

17 For details, see Northrup, *Boulwarism, op. cit.*, pp. 14-19, 39-44; and Walter Galenson, *The CIO Challenge to the AFL: A History of the American Labor Movement, 1935-1941* (Cambridge: Harvard University Press, 1960), pp. 239-265.

18 An explanation and discussion of "Boulwarism" by a former practitioner is found in Northrup's book of the same name, *op. cit.*

At the national or international level, the policies of the unions in matters of civil rights have been varied. The IBEW and the IAM, consistent with the older craft union tradition, have been conservative. Like some other unions, the IAM, prior to 1949, formally excluded Negroes from membership by a provision in its ritual. Today, this union pursues a "laissez-faire" policy, leaving most racial policies to local union determination, and neither opposing Negro employment nor giving significant support to affirmative action to employ Negroes.

The IBEW never barred Negroes by formal provision, but did so by common consent. After it began organizing factory workers, the IBEW accepted Negro members, but many of its craft locals, particularly in the construction industry, have vigorously opposed integration, and some undoubtedly still do. In general, as a craft-oriented union like the IAM, the IBEW has shown only moderate enthusiasm for changes in racial practices.

The UE and the IUE, with origins in the CIO, are more ideological in their orientation and, at the national level, much more vigorous in their support of social change. For many years, James B. Carey, former president of the IUE, was chairman of the CIO Civil Rights Committee and, given the limited power of this committee, was active in support of increased Negro employment. The international IUE has taken a strong stand on civil rights, conducting vigorous social action and conferences. President Paul Jennings was cited in 1967 by the National Urban League for "outstanding contributions to equal opportunity."

The formal posture of the electrical unions, at both the international and local levels, is expressed in the typical nondiscrimination clause in collective bargaining agreements:

> The Union, the Locals, and the Company reaffirm their intention that the provisions of this Agreement and of local supplements will continue to be applied without discrimination because of race, creed, color, sex, or national origin of the employee.[19]

Commitment to implementing policy, of course, depends on the membership. Since control by the internationals has not generally been vigorous in the electrical industry, the attitudes and practices of the local unions depend primarily on the members and leaders of the locals themselves.

---

[19] Westinghouse agreement with the IUE, October 17, 1966, p. 3.

Electrical local unions have not often officially opposed integration, but a few locals, for example in Chicago and the Pittsburgh area, have opposed job training programs for ghetto blacks, calling this preferential treatment. Most locals have been reluctant to modify seniority clauses that slow down the advancement of low seniority blacks, but those clauses are not *per se* discriminatory. On the other hand, locals with a high percentage of black members have pushed vigorously for upgrading blacks. Locals have often been slow to admit Negroes into positions of union leadership.

We should add here that the nature of industrial unionism mitigates the influence of the union in racial policy and practice. Management, not the union, controls the hiring policy, thereby establishing the racial composition of the work force and, indirectly, of the local union. Union membership does not precede but follows employment. On balance, unions in the electrical industry have served neither to inhibit nor to advance minority employment policies.

REGION AND LOCATION

The electrical industry is located throughout the United States with heaviest concentration in the Northeast and Midwest. Table 8 shows that 32.4 percent of electrical employment is in the Northeast (New England and Middle Atlantic regions), primarily in New York, Pennsylvania, New Jersey, and Massachusetts. The Midwest accounts for 32.8 percent of electrical manufacturing employment with Illinois, Ohio, and Indiana employing by far the greatest numbers. Midwest employment ranges from 179.1 thousand in Illinois to 3.9 thousand in Kansas. The South has 19.4 percent of electrical employees, the Rocky Mountain region only 2.4 percent. The West Coast accounts for 13 percent of total electrical employment with 214.6 thousand of the 226.6 thousand employed in California.

Older electrical plants were built in small manufacturing cities and towns in the Northeast, many of which still have very small Negro populations. Schenectady, New York, Lynn and Pittsfield, Massachusetts, and Erie, Pennsylvania remain major centers of the industry today and have less than 5 percent blacks in the labor market. Lowell, Utica, Ashland, Massachusetts, and many other smaller towns also have very small black populations.

TABLE 8. *Electrical Manufacturing Industry Employment by Region United States, 1969*

	All Employees		White Collar Employees		
Region	Total	Percent Distribution	Total	Percent Distribution	Percent of Total
New England	138,737	8.0	55,504	8.2	40.0
Middle Atlantic	426,882	24.4	177,585	26.3	41.6
Midwest (North Central)	571,966	32.8	174,407	25.9	30.5
South	339,329	19.4	121,132	18.0	35.7
Mountain	42,576	2.4	17,433	2.6	40.9
West Coast (Pacific)	226,609	13.0	127,881	19.0	56.4
Total[a]	1,746,099	100.0	673,942	100.0	38.6

Source: Preliminary Equal Employment Opportunity Commission data, 1969.

[a] Total is sum of regions. See note, Appendix Table A-4 regarding use of EEOC United States and regional totals.

Definition of Regions:

New England: Connecticut, Maine, Massachusetts, New Hampshire, Rhode Island, Vermont.

Middle Atlantic: New Jersey, New York, Pennsylvania.

Midwest (North Central): Illinois, Indiana, Iowa, Kansas, Michigan, Minnesota, Missouri, Nebraska, Ohio, South Dakota, Wisconsin.

South: Alabama, Arkansas, Delaware, District of Columbia, Florida, Georgia, Kentucky, Louisiana, Maryland, Mississippi, North Carolina, Oklahoma, South Carolina, Tennessee, Texas, Virginia, West Virginia.

Mountain: Arizona, Colorado, Nevada, New Mexico, Utah.

West Coast (Pacific): California, Oregon, Washington.

Other plants, originally built in white metropolitan neighborhoods, have been affected by Negro migration, especially in northern cities such as Philadelphia, Trenton, and Newark, where 30 percent or more of the labor force is black. Companies with such older plants do not need to start new "ghetto industries." They already have such industries. Their problem in the 1970's will be to get qualified blacks to man their plants. They will probably find it difficult to attract whites into such a work force, but these plants will provide easily accessible jobs for many blacks.

The number of blacks in the labor force near midwestern plants also varies greatly. Chicago, Cleveland, Cincinnati, Columbus, Detroit, and Louisville—all major electrical centers—all have very large Negro populations. On the other hand, Fort Wayne, Indiana, with several large electrical machinery plants has a smaller percentage and smaller centers like St. Joe, Michigan, Mattoon, Illinois, and Bellevue, Ohio have very few Negroes.

The West Coast has a very high concentration of professional and technical employees. Although only 13 percent of all electrical employment is in the Pacific region, 19 percent of all white collar employment is there, and white collar employees make up 56.4 percent of the work force (Table 8). In white collar work the absolute number of Negroes in the labor market is not nearly so important as the educational level of the blacks available. Large electrical manufacturing plants are located mostly near Los Angeles and Oakland each with a large Negro population.

Many new plants are being built along major highways around metropolitan areas, such as Route 128 and Route 495 near Boston, restricting employment possibilities for unskilled and semiskilled black workers in the central cities. The poor are least mobile for reasons both economic and psychological. Public transportation is often slow, complicated, and expensive. Relocation close to the suburban work site may be possible for whites, but very difficult for blacks because of housing restrictions, a threatening social climate, or simply unwillingness to leave the familiar black community. The result: jobless Negroes in the cities and expanding, predominantly white plants outside the cities.

Southern plants which are relatively new are often located in areas more accessible to whites than to blacks because of transportation problems, but almost all are near large numbers of potential black employees. As we have noted, most new plants are located in suburban or rural areas, in pleasant surroundings, but without access to public transportation. This tends to mitigate against the employment of the disadvantaged, many of whom are Negroes.

The electrical companies also have a great number of service and sales districts in towns and cities throughout the United States. Many are in central city areas with high black concentrations in the labor force. In fact, there are probably small offices like these near every concentration of Negroes in the country. The problem, as we shall point out, is that it is difficult to find very many Negroes for electrical servicing and sales jobs.

SOME TENTATIVE HYPOTHESES

We have estimated that the expansion of the industry will result in 45,000 new jobs per year over the next decade and that normal turnover could provide 4,700 more for a total of 49,700. A very crude benchmark would provide that if 10 percent of these jobs went to Negroes, it would mean nearly 5,000 jobs per year for black people.

Much more than just numbers, however, is involved. For example, the impact of the occupational structure of the industry must be considered in the light of its sizeable technical and professional needs. To put the possible impact of the electrical manufacturing industry's occupational distribution in proper focus, Table 9 compares it with that of six other industries—automobiles, aerospace, steel, rubber tires, petroleum, and chemicals—in 1966. It is immediately clear that electrical machinery's occupational distribution ranks in the middle between the overwhelmingly blue collar industries such as automobiles, steel, and rubber tires and the heavily white collar employers such as aerospace and petroleum. The electrical industry distribution is quite similar to that of chemicals. The big difference is that the latter has almost twice the proportion of craftsmen and a considerably smaller proportion of operatives. In comparing the six industries which, besides electrical manufacturing, are represented in Table 9, Northrup notes:

> Given this occupational mix and the disadvantaged status of Negroes . . . insofar as educational attainment, training, and industrial experience are concerned, one would expect that the proportion of Negroes would be highest in automobiles, steel, and rubber tires, lowest in aerospace and petroleum, and chemicals above the later two, but below the first three. Of course, this is precisely the case . . . .[20]

Following this same reasoning, we would expect that the proportion of Negroes in electrical manufacturing will be near to, but above that of chemicals because of the similar occupational distribution but greater proportion of lower skilled jobs in the former, and that it will be below that of automobiles, steel, or rubber tires. This would not result in 10 percent accretion of new employees.

There are other reasons why a 10 percent share of new jobs may not go to Negroes. For this to occur, new Negro employees

---

[20] Northrup, Rowan, *et al., Negro Employment in Basic Industry,* Part Eight, p. 724.

TABLE 9. *Seven Manufacturing Industries*
Percent Distribution of Employees by Occupational Group
1966

Occupational Group	Automobiles	Aerospace	Steel	Rubber Tires	Petroleum	Chemical	Electrical Machinery
Officials and managers	7.9	9.1	8.3	8.5	10.5	12.1	6.9
Professionals	5.4	22.8	2.4	5.1	13.4	10.5	10.3
Technicians	2.6	8.1	2.3	3.0	5.0	6.8	6.6
Sales workers	—	0.1	0.6	2.1	10.8	1.9	1.3
Office and clerical	9.8[a]	16.5	8.6	9.5	19.4	11.1	11.9
Total white collar	25.7	56.6	22.2	28.2	59.1	42.4	37.0
Craftsmen	13.6	20.9	23.9	14.8	18.5	20.7	12.0
Operatives	54.5	19.7	37.1	49.7	17.5	29.2	40.4
Laborers	3.4	1.0	15.0	4.1	3.5	5.0	9.1
Service workers	2.8	1.8	1.8	3.2	1.4	2.7	1.5
Total blue collar	74.3	43.4	77.8	71.8	40.9	57.6	63.0
Total	100.0	100.0	100.0	100.0	100.0	100.0	100.0

Source: Computed from Appendix Table A-2; and Herbert R. Northrup, Richard L. Rowan, *et al.*, *Negro Employment in Basic Industry*, Studies of Negro Employment, Vol. I (Philadelphia: Industrial Research Unit, Wharton School of Finance and Commerce, University of Pennsylvania, 1970), Part Eight, p. 724.

[a] Includes the few classified as sales in the industry.

would probably have to have the same turnover rates as existing workers, while turnover is generally higher among the disadvantaged and the new recruits to industry.

On the other hand, much may be determined by the character of management action. If traditional employment practices and conventional labor sources are utilized, hiring of Negroes will fall much below this benchmark; but if "affirmative action" policies induce corporate personnel officers to break new grounds in recruiting and training, new Negro employment may exceed even reasonable expectations.

Let us speculate briefly on a related point: the electrical manufacturing industry has been characterized not only by rapid growth but also by change. Growth is not merely quantitative but technological. Managers in the electrical industry have been receptive to new methods of marketing and production. They have been concerned about new plants and new locations and, indeed, new definitions of their own firms and their own industry. They plan to extablish hundreds of new plants in the seventies.

Does this disposition to change have transfer value? Will a management that is congenial to technological change also have an openness to change in the political and sociological dimensions of industry and economic society?

Here we merely express a hypothesis: namely, management which is ready for alterations in product lines, productive processes, and in marketing procedures will be equally equipped to apprehend and implement changes in manpower planning and practice. To extend the hypothesis: in an industry such as electrical manufacturing, one may expect to find a management pool of talent and flexibility conditioned to change and able to understand and direct the process of change, whether change is technological or institutional.

There is another reason why we might expect managerial policies to increase Negro employment. Much of the industry is consumer oriented. As Northrup has noted:

> Companies which sell to consumers usually align their marketing and racial policies very carefully. They are more likely to be on the lookout for means to avoid offending any group, and especially majority opinion. Thus, they desire to move slightly ahead, but not too far. It is not unusual for such companies to work assiduously behind the scenes in a community to work out problems. Racial unrest can alienate a group, alter buying habits, and is generally bad for business.[21]

[21] *Ibid.*, Part Eight, p. 734.

Electrical manufacturers want to sell appliances, radios, television sets, and other consumer goods to *all* customers. The Negro market is well worth cultivating, and it can readily be done with little or no backlash. As large defense producers, moreover, electrical manufacturers strive to conform to government procurement policies.

All this adds up to the belief that both the electrical manufacturing industry's character and direct economic interests would lead it to accept a substantial increase in the employment and upgrading of Negroes as a wise challenge well within the competence of the industry's management. Since the industry has relatively recently recognized this challenge, the hypothesis cannot be fully tested until progress made in the 1970's can be evaluated. There are signs that the hypothesis has some validity, such as General Electric's new Business Environment Service, which forecasts the company's social and political world in the 1970's, and GE's use of the systems approach to minority manpower planning.

We shall evaluate the situation to date in this study. First, it is necessary to examine the history of Negro electrical manufacturing employment.

CHAPTER III

Negro Electrical Manufacturing Employment to 1960

Negro employment in the electrical manufacturing industry is largely a phenomenon dating from World War II. Prior to then, less than one percent of the industry's employment was black. Why this was so, and what factors induced change, are discussed in this chapter.

THE PERIOD PRIOR TO WORLD WAR II

The electrical manufacturing industry began in the 1870's when electric arc lights and power generators first came into very limited commercial production. The decades following saw rapid technological advance and notable expansion in production. By 1890 Western Electric was already over twenty years old, and Westinghouse had been founded. In 1891 General Electric, a consolidation of predecessor companies, came into existence.

Employment in the industry increased rapidly. At the end of the nineteenth century, the industry employed almost 50,000 workers. By 1910 the figure was over 100,000, by the end of World War I, over 300,000. In 1929, just before the Depression, 420,000 workers were employed, or 4.4 percent of the total employment in American manufacturing.[22] There were very few Negroes among the growing number of electrical employees. In 1930 less than one percent of all employees were black, as shown in Table 10.

One major reason for the absence of Negroes was the location of the industry. Electrical manufacturing was a northern industry, concentrated in the Northeast. Moreover, manufacturing locations were mostly in the smaller manufacturing cities. In 1910 less than 2 percent of the population of the northeastern and north central states was Negro, and a good part of this num-

[22] All data are from Backman, *The Economics of the Electrical Machinery Industry*, p. 328.

TABLE 10. *Electrical Manufacturing Industry Employment by Race and Sex United States, 1930-1960*

Year	All Employees			Male			Female		
	Total	Negro	Percent Negro	Total	Negro	Percent Negro	Total	Negro	Percent Negro
1930[a]	383,570	3,095	0.8	298,383	2,919	1.0	85,187	176	0.2
1940[b]	374,684	1,782	0.5	273,483	1,683	0.6	101,201	99	0.1
1950[b]	788,184	19,433	2.5	512,760	11,651	2.3	275,424	7,782	2.8
1960[b]	1,487,412	48,903	3.3	978,259	30,730	3.1	509,153	18,173	3.6

Source: *U.S. Census of Population:*

1930: Vol. V, Occupations—General Report, Chapter VII, Table 2.

1940: Vol. III, The Labor Force, Part 1, Table 76.

1950: Vol. II, Characteristics of Population, Part 1, Table 133.

1960: PC(1)ID, U.S. Summary, Detailed Characteristics, Table 213.

Note: It is believed that census data underestimate total employment in the industry, but there is no reason to believe that the proportion of Negroes is disproportionately represented. Compare Table 4, which shows the trend of employment as determined by the U.S. Bureau of Labor Statistics from payroll data.

[a] Gainful workers (10 years and older).

[b] Employed persons (14 years and older).

ber was concentrated in the largest cities of the Northeast. Negroes were hardly present in the labor pool from which the electrical industry drew its work force. Even as late as 1940 only 3.8 percent of the population in the Northeast and North Central states was black.[23]

During World War I, the electrical industry was not subject to some of the pressures which led such industries as automobiles and steel to recruit large numbers of southern Negroes, or even to do the small amount of such recruiting done by the rubber tire industry in this period.[24] Its labor requirements at that time were more heavily skill oriented than many other industries and for such work; the first and second generation immigrants and older American recruits from rural areas proved adequate. Where unskilled labor was needed, such as lamp (bulb) manufacturing, white women whose husbands were often working in steel or metal plants proved sufficient.

With the industry emphasizing good benefits and stable or expanding employment, it did not become involved in labor struggles involving the use of strikebreakers. Hence there was no importation of black strikebreakers as in steel or meat packing.[25] Finally, unlike heavy industries such as smelting and steel, or industries with less desirable work such as meat packing, electrical manufacturing did not have the concentration of arduous or hazardous jobs that so often became known as "Negro jobs."

Early Management Policy

That Negroes were not considered for the labor force of this industry in these years is obvious from Table 10. In 1930, only 0.8 percent of the work force was black. Moreover, during the Depression, the tiny Negro work force was further depleted, declining at a more rapid rate than did total employment during these difficult years. Hence, in 1940, on the eve of World War II, only 0.5 percent of the employment was black.

---

[23] U.S. Bureau of Labor Statistics, *The Negroes in the United States: Their Economic and Social Situation*, Bulletin 1511 (June 1966), p. 66.

[24] See Northrup, Rowan, *et al.*, *Negro Employment in Basic Industry*, Part Two, pp. 51-53; Part Four, pp. 400-402; and Herbert R. Northrup, *Organized Labor and the Negro* (New York: Harper and Brothers, 1944), pp. 172-174.

[25] Northrup, *Organized Labor and the Negro, loc. cit.;* Theodore V. Purcell, *The Worker Speaks His Mind* (Cambridge: Harvard University Press, 1953), pp. 46, 51; and Walter A. Fogel, *The Negro in the Meat Industry*, in the Racial Policies of American Industry series, pp. 30-38.

Despite the small number of Negroes in the industry, management of large firms became aware of the problem at an early date. The first incident occurred during World War I. General Electric's Schenectady works was unionized by the "Electrical Industry Trades Alliance," a confederation of craft unions affiliated with the American Federation of Labor and chartered by its Metal Trades Department. The Alliance (later termed Metal Trades Council) "and its constituent unions bargained for their members aggressively and with considerable success."[26] In June 1917, one of the unions involved, the International Association of Machinists, which then excluded Negroes, struck General Electric because one of thirty college students employed for summer work was a Negro, Wendell King. The IAM demanded the immediate dismissal of King from the machine shop. When management flatly refused to accede to their deman, an eight day strike occurred which was the first in the history of the company revolving around the race issue. It was clearly a Machinists' affair; no other labor unions supported it.

The following statement, issued by the company, describes management's stand in the case, as well as the fears of the union Machinists:

> It is not the intention of the company to introduce colored labor into the shops or displace any white labor with colored labor, nor is it the intention of the management to bring numbers of colored people to Schenectady to work in the shops, and we recommend that any objection to the young colored man, Wendell King, remaining at work in the department where he is now employed be withdrawn, and that the Machinists now on strike return to their places and resume work considering the matter at a closed issue.[27]

The strike ended on company terms. There is no record of further racial strife or of company action to increase its Negro employment. Yet in 1935, General Electric issued a booklet entitled "GE Policies Concerning Wages, Hours, and Working Conditions of Shop Employees." Included was this statement:

> There shall be no discrimination by foremen, superintendents, or any executives of the Company against any employee because of race, or creed, or because of an employee's membership in any fraternity, society, labor organization or other lawful organization.

---

26 Milton Derber, "Electrical Products," in Harry A. Millis, *et al.*, *How Collective Bargaining Works* (New York: The Twentieth Century Fund, Inc., 1942), p. 748.

27 From newspaper accounts of the incident on file at General Electric Company. See, e.g., *Schenectady Union Star*, June 21, 22, 25, and 26, 1917.

The booklet was signed by President Gerard T. Swope, whose unconventional (for the period) stance in regard to unions has already been noted. Why the statement was issued is not clear, but Swope's deep concern about social problems may have been the reason. In any case, it does appear to have been the first such policy announced in writing by a large company.[28]

General Electric's management was not alone in the industry in recognizing the race problem before 1940. RCA cites a written merit employment policy of 1919;[29] and "Western Electric, manufacturing and supply unit of the Bell system, has had a policy of recruiting, hiring and assigning employees on the basis of merit alone at least since 1924, when it was spelled out in a document defining employment policies."[30] Unlike the later General Electric policy statement, the latter two did not specifically mention "race or creed."

These statements indicated commitment to an ideal and they established a goal which was later implemented, although practice at the time fell far short. RCA, like most electrical manufacturers, did not employ Negroes in production or clerical jobs before 1941,[31] and integration at the Kearny, New Jersey, Western Electric plant in the same year was considered a breakthrough. General Electric followed up its policy statement of 1935 by including a nondiscrimination clause in its contract with the United Electrical, Radio and Machine Workers, yet sizeable integration progress had to await the war years. As late as 1941, the U.S. Bureau of Employment Security reported its findings that managers surveyed in the electrical manufacturing industry would *not* consider a Negro, if available for 10,346 of the 20,792

---

[28] See General Electric Company, *Community and Government Relations* Bulletin 64-3 (April 17, 1964); and Virgil B. Day, "Progress in Equal Employment Opportunity at General Electric," in Herbert R. Northrup and Richard L. Rowan (eds.), *The Negro and Employment Opportunity* (Ann Arbor: Bureau of Industrial Relations, University of Michigan, 1965), p. 155. For an account of Swope's remarkable career, including his interest in social work and social problems, see Loth, *Swope of GE.*

[29] Frank M. Folsom, President, RCA, statement on "Equal Opportunity in Employment," before the United States Senate Committee on Labor and Public Welfare, Subcommittee on Civil Rights, February 23, 1954, reprinted by RCA, pp. 5 and 6.

[30] R. G. Lawrence, "Western Electric's Commitment to Fair Employment," in Northrup and Rowan, (eds.), *The Negro and Employment Opportunity,* p. 137.

[31] See Mr. Folsom's statement before the Senate subcommittee, *loc. cit.*

job openings which they anticipated would become available within the next several months.[32]

Certain skilled jobs in some plants were undoubtedly closed because of contracts with the International Association of Machinists (IAM), the International Brotherhood of Electrical Workers (IBEW), or other exclusive craft unions, but these unions represented only a small proportion of electrical workers at a few locations. More likely management was reluctant to face the anticipated reaction of white workers.

In fact, management rarely had to consider the issue. Even where locational factors or inadequate educational background did not rule out potential Negro job applicants, it is likely that few Negroes applied because they believed that it would not be a fruitful exercise. In 1940, the electrical industry was virtually all white. The very few Negroes employed were generally excluded from production jobs. Despite fair employment policies, managements seem to have accepted the mores of the time, which limited Negro employment opportunity.

WORLD WAR II AND ITS AFTERMATH

World War II caused a rapid and massive increase in the demand for labor which brought Negroes into the electrical industry in substantial numbers for the first time. In 1940 the effects of the Depression were still evident. The unemployment rate was close to 15 percent with 8 million unemployed. In the four subsequent war years, the armed forces increased from one-half million to over eleven million. At the same time, American manufacturing expanded to meet war production and 7.3 million additional people were employed. Unemployment dropped to a historic low of 1.2 percent in 1944. Inevitably the composition of the American labor force changed substantially.

The electrical manufacturing industry was particularly affected by the wartime demand for labor since employment expanded to 1,087,000 in 1944, more than double what it had been in 1940. (See Table 4, Chapter II.) The manpower pressures of massive war production also produced changes in job content and description which favored low skilled, poorly educated workers. Production systems were redesigned and many low skilled, repetitive jobs were created. New electronic products and systems which were developed in this period required large numbers of low

[32] "The Negro's War," *Fortune*, Vol. XXV (June 1942), p. 9.

skilled assembly operations. Lower productivity was tolerated, and government subsidies were available for recruiting and training inexperienced workers.

In addition to changes in the labor market, there was a militant Negro drive for an end to job discrimination under the leadership of A. Phillip Randolph. A demonstration march on Washington was threatened, and on June 25, 1941, President Franklin D. Roosevelt issued Executive Order 8802 announcing that discriminatory practices by government and defense contractors were contrary to the public interest. This Executive Order also established a President's Committee on Fair Employment Practice. David Sarnoff, president of RCA, was an original member of this committee, but he resigned after the Committee's projected railroad hearings were postponed, alleging that the Committee had proved ineffectual. A second Executive Order, No. 9346, issued on May 27, 1943, with a more vigorous approach, dramatized the need for nondiscrimination and contributed to fair employment in a number of industries despite its lack of enforcement machinery. Although the existence of FEPC did give employers desirous of adding Negroes to the work force support against recalcitrant management colleagues and fearful or disgruntled employees, the labor shortage was nevertheless, the primary impetus for employing Negroes.

The early 1940's saw significant breakthroughs in the number of Negroes hired and some progress in the kinds of jobs for which Negroes were employed. Among the first to inaugurate change in the electrical manufacturing industry was Western Electric in its Kearny, New Jersey plant, which employed about 2,000 Negroes by the end of the war. *Fortune* gives this euphoric account:

> A year ago [in 1941] Western Electric hired Negroes for the first time in its Kearny, N.J. plant and put them to work side by side with skilled white people. The experiment was a full success. No separate cafeterias, no separate washrooms, and no move to get them. The plant is producing highly complicated instruments; neither the work nor its quality was even slight disturbed.[33]

Often management met some resistance to integration of production jobs, but found that its fears of white reaction were exaggerated. Thus when RCA recruited Negroes for production jobs in Indianapolis in 1941 and in Camden and Harrison,

[33] *Ibid.*, p. 158.

New Jersey somewhat later, it was necessary to dismiss "only a very few whites who refused to work with blacks." [34] Westinghouse had more trouble at the headquarters plant in East Pittsburgh when the company hired black men for the production lines; and rest room facilities were vandalized when black women originally hired on third shift gained enough seniority to move up to the same shift with white women. The United Electrical, Radio and Machine Workers' local union supported management in both cases, and blacks and whites were soon working side by side without further incident.[35] At General Electric's large Schenectady plant, scene of a strike against the employment of a Negro college student during World War I, large numbers of Negroes were employed without serious difficulties, and the company was cited for its progressive racial policies.[36]

It is difficult to estimate the number and proportion of Negroes employed during the war, since no data are available. It is unlikely, however, that the proportion exceeded 3 percent, and probably was closer to 2.5 percent, the proportion found by the 1950 census. This would mean that about 30,000 Negroes had jobs in the industry in 1944.[37]

By the end of the war, barriers remained in the industry. Negroes were rarely assigned to white collar or supervisory jobs. The significant factor, however, was that Negroes had been employed for the first time in the industry in large numbers and were working along with whites. Moreover, the great migrations during the war for the first time brought Negroes within the potential work force of many companies, and therefore offered the opportunity for future black additions to the industry's employment roles.

Postwar Readjustment

Although the electrical manufacturing industry declined by approximately 150,000 from its wartime peak in the first postwar

---

[34] Statement of Frank M. Folsom, *loc. cit.*

[35] Field notes and interviews.

[36] John A. Davis, "Negro Employment: A Progress Report," *Fortune*, Vol. XLVI (July 1952), pp. 102, 103, 158-162.

[37] This is approximately 2.8 percent of total employment of 1,087,000 in 1944. (See Table 4). Census data in Table 10 appear to underestimate employment.

year, it recovered rapidly. Both its products for industry and for the consumer had enormous backlogs of orders and demand. In addition, the late forties saw the first stages of the burgeoning electronic revolution. By 1947 electrical employment was approximately equal to its wartime high. Even in 1949, a low point in postwar industry employment, the electrical work force at 862,000 was about 80 percent of its maximum wartime employment and almost twice its 1940 level; by 1951, employment had passed the wartime peak.[38]

Whether Negroes were affected disproportionately by the postwar layoffs cannot be determined. In any case, in 1950, when employment in the industry was still below the wartime peak, Negro employment, as reported by the U.S. Census, was 2.5 percent that of total employment reflecting an absolute increase of over 17,500 jobs for Negroes (Table 10). Thus in the decade of the 1940's, although Negro employment increased faster than total employment, still because of the low base of 1940, Negroes by 1950 represented only a small proportion of the industry's labor force.

Integration Activities

During the war years electrical manufacturers put into practice previously held policies regarding employment of Negroes. Manpower needs dictated that they employ Negroes and nearly all such employment was confined to blue collar jobs. The postwar period saw the development of the first programs designed to employ Negroes as a result of a concern for integration. Moreover, these programs were designed to obtain positions for other than blue collar personnel.

In the early part of 1947, the National Urban League established a Management Advisory Committee to place competent blacks in professional and technical positions. A. B. Goetze, then Personnel Director of Western Electric, and Dr. G. Roy Fugal, long General Electric's equal employment coordinator, were active in initiating the formation of this Committee. Moreover, in succeeding years, representatives of a number of electrical manufacturers were active on the Committee, including Westinghouse and RCA, and began consulting regularly with the National Urban League on the corporate level.

---

[38] See Table 4 and U.S. Bureau of Labor Statistics, *op. cit.* p. 302.

Also in 1947, Western Electric began an executive training program dealing with equal employment opportunities. Some firms commenced informal audits of their own Negro employment, with tentative steps to increase it. Even though the commitment of top managers often was not effectively implemented; jobs did result, token at first, but gradually increasing for Negroes with key skills.[39]

In 1948, Howard University became the first predominantly Negro university to have its engineering school accredited by the Engineer's Council on Professional Development. The following year General Electric, at the suggestion of the National Urban League, became the first nationwide company to include the Howard engineering school in its regular college recruiting. Other electrical machinery concerns, as well as other companies, soon followed suit in *actively* recruiting black technical personnel.

Consultation with the Urban League had little effect on the overall racial composition of the industry. Efforts were directed primarily toward a few individual pioneering breakthroughs at professional and technical levels, rather than toward large numbers of Negro recruits. Although fair employment programs were not a high priority, these efforts were the first steps toward voluntary elimination of the strong, though unwritten, barriers which excluded Negroes from many jobs in the industry.

Other developments in the late 1940's also showed a concern for minority employment. In 1946-1947, Charles E. Wilson, Swope's successor as President of General Electric, served as chairman of the President's Committee on Civil Rights, appointed by President Harry S. Truman. The Committee called for an end of all segregation and discrimination and the passage of effective civil rights and equal employment opportunity legislation. Wilson who had reissued the Swope nondiscrimination policy in 1941, signed the cover letter transmitting the report to President Truman stating in part

> the protection of Civil Rights is a national problem which affects everyone. We need to guarantee the same rights to every person regardless of who he is, where he lives, or what his racial, religious or national origins are.[40]

---

[39] Much of this information is based on interviews with Dr. Julius Thomas, for many years Industrial Relations Secretary of the National Urban League, who began the programs described here.

[40] Copy in authors' possession.

This was perhaps the first call for such legislation by a major industrialist.

In the late 1940's also, seven additional states—New Jersey, Massachusetts, Connecticut, New Mexico, Oregon, Rhode Island, and Washington—followed New York in passing state fair employment practice legislation with enforcement provisions. Thus in the northeastern states, where the electrical manufacturing industry was traditionally concentrated, equal employment opportunity had become a matter of law by 1950. In the South, however, where the industry was beginning to expand, segregation was still legal and enforceable.

THE DECADE OF THE 1950's

In the electrical manufacturing industry the decade between 1950 and 1960 was marked by uneven but substantial growth and continued improvement in the employment of Negroes. The Korean War (1950-1953) again drew civilian workers into the armed forces and increased production and total manufacturing employment. Employment in the electrical manufacturing industry reached 1.3 million in 1953, an increase of approximately 400,000 new employees since 1950, a three-year period when the national unemployment rate was dropping from 5.3 to 2.9 percent. Increased demand for labor in a tight labor market again meant increased opportunities for Negroes.

Since it was a growth industry, electrical manufacturing was not so severely affected as other industries by post-Korean War readjustment. Although the electrical industry lost 143,000 jobs in 1954, it recovered rather well and employed 11,000 more persons in 1957 than in the Korean War peak year of 1953. The recession year of 1958 cost the industry 95,000 jobs, but by 1960, industry employment stood at 1,467,000, a gain of 134,000 over 1953 and 476,000 more than in 1950. (See Table 4, Chapter II.)

Within the industry, however, a number of companies felt the 1958 recession more sharply and for a longer period. General Electric employment, for example, fell by more than 50,000 from 285,000 in 1957 to 231,000 in 1959, and did not regain the 1957 figure until 1966.[41] Other companies suffered similar cutbacks, but the industry as a whole increased employment in the 1950's

[41] Data from company annual reports.

at about four times the rate for manufacturing industry generally.

As in industry generally, employment of salaried personnel increased at a faster rate in the 1950's than did that of blue collar workers. The development of missiles and sophisticated electronic equipment and controls required a high engineering content. General Electric, which supports a vast research program, saw its salaried employment exceed its hourly workers in the late 1950's.

During the 1950's, emphasis on automation and methods improvement in American industry frequently meant the elimination of unskilled jobs and with that, a decline in the proportion of Negroes, who were concentrated in the unskilled jobs and either denied opportunities, or by reason of inferior educational opportunities, found unqualified for positions of greater skill. Chemicals, paper, and tobacco are examples of industries in which the impact of technology resulted in a net decrease in the proportion of Negroes during this decade.[42]

In electrical manufacturing, however, production jobs continued to expand, so that 226,000 more such workers were employed in 1960 than in 1950 (Table 4). Assembly workers for appliances, electronic controls and components, televisions, and other such jobs required an ever larger number of semiskilled employees. The greater emphasis on civil rights and the increasing availability of jobs of all kinds resulted in continued improvement in the proportion of Negroes in the industry. Table 10 shows that by 1960, 3.3 percent of the employees were black, as compared with 2.5 percent in 1950.

Plant Location and Southern Expansion

It will be recalled that locational factors before World War II reduced Negro participation in the industry since few Negroes then lived near the plant concentrations. By end of the war, the great migrations from southern rural areas to the North had altered population patterns, but as the electrical machinery industry began building new plants, Negroes again found them-

---

[42] See Northrup, Rowan, *et al., Negro Employment in Basic Industry,* Part Seven, pp. 623, 628-629; and Herbert R. Northrup, Richard L. Rowan, *et al., Negro Employment in Southern Industry,* Studies of Negro Employment, Vol. IV (Philadelphia: Industrial Research Unit, Wharton School of Finance and Commerce, University of Pennsylvania, 1971), Part One, pp. 34, 47-48, and Part Three, pp. 30-33.

selves at a disadvantage. Many of these new plants were constructed in the more rural or suburban areas of the North and West where the labor force was overwhelmingly white.

In addition, new plants in urban areas were located outside of city limits so that companies would be able to construct single story structures with ample room for expansion and thus avoid the higher taxes, increased material handling costs, and lack of expansion potential which city location usually involves. More and more such plants have been built in areas completely dependent on automobile transportation for employees. As public urban transportation steadily declined throughout the 1950's, jobs became more dependent on car ownership. Added to the locational problem which Negroes who live in center city have in obtaining jobs in outlying plants, there is the transportation problem, since proportionately fewer Negroes own automobiles than do white.

The late 1940's and most of the 1950's saw a major expansion of the electrical machinery industry into the South. Most of the plants built there were located in small towns or on the outskirts of cities. In nearly all cases, they could be reached only by automobile, thus limiting Negro access and employment. In addition, jobs in these plants were much in demand. Electrical machinery employers, like those in other industries, were able to set high hiring standards. Discriminatory education ensured that proportionately fewer Negroes could meet the standards.

Negro employment was also hindered by the discriminatory policies of southern state employment services. Many companies had their prespective employees screened by state employment officies, which often did not refer Negroes to good jobs. In some states laws requiring separate washroom and eating facilities for black employees made hiring Negroes very expensive. An Urban League survey of one major electrical corporation in 1956 indicated that only 2.6 percent of its employees in the southern and border states were black despite the much larger number of Negroes in most southern areas.[43]

Early in the decade, most management accepted southern racial mores except where individual managers were more concerned about the situation. Thus General Electric Company, in accordance with its newly developed decentralized organization, appears

---

[43] Copy in authors' possession.

to have left racial policies to individual managers. As a result, at least three of its early post-World War II plants employed a lily-white work force and a few others hired Negroes only in unskilled jobs, but at least four others made significant attempts to employ Negroes on an integrated basis.[44]

General Electric's integrated plants were the results of individual managerial actions to which top executives gave support. Thus when white workers walked off the job at the Louisville, Kentucky "Appliance Park" complex, because Negroes were employed on the assembly line, the strikers were discharged or otherwise disciplined and the integration efforts continued.[45]

One General Electric executive, Dr. Louis Rader, not only pushed integration at the plant under his jurisdiction (Waynesboro, Virginia), but also publicly helped to lead the fight in that state which reopened the schools and ended Virginia's ill-fated "massive resistance" to school integration. Dr. Rader's position, like that of other industrialists so committed, was simple: without good public schools, industry—and especially highly technical industry—could not attract engineers and other professionals to live and to work in the South. The choice was between following the mores of the Old South, or moving with industry.[46]

The pattern at Westinghouse was similar. The company generally deviated only slightly from local southern custom. One plant in the deep South had a 6 percent black work force but confined the black men to unskilled jobs. Minor modifications were made in nonsegregated eating and washroom facilities in some states. Yet the Baltimore plants, opened in the early fifties, employed Negroes in all capacities contrary to local custom and even held company parties in a federal armory because racially mixed parties were not allowed in local hotels.[47]

---

[44] Interview with a former General Electric employee relations executive, February 2, 1970.

[45] *Ibid.*

[46] For an account of Dr. Rader's efforts and those of executives of other companies, see Herbert R. Northrup, "Industry's Racial Employment Policies," in Arthur M. Ross and Herbert Hill (eds.), *Employment, Race, and Poverty* (New York: Harcourt, Brace & World, Inc., 1967), pp. 294-295; and William H. Nicholls, *Southern Tradition and Regional Progress* (Chapel Hill: University of North Carolina Press, 1960), pp. 114-123.

[47] Remarks by Gwilym A. Price, President and Chairman, Westinghouse Electric Corporation, The President's Committee on Government Contracts, Conference of October 25, 1955, reprint from Westinghouse.

The genuine concern of industry managers with the threat posed to effective operations and personnel recruitment by the southern school crisis of the 1950's led many to take a more searching look at their plant personnel racial policies. At the very least, a greater receptivity resulted for governmental integration efforts. Moreover, the disadvantages resulting from the furor over schools made top management more conscious of southern race problems and less willing to believe that adherence to old southern employment mores could ensure tranquillity or acceptance.

Even in the border or northern states, decentralized management unconcerned with corporate equal employment programs produced results inconsistent with corporate policy. One large company's headquarters office personnel recruiter used several New York City parochial schools as the principal source of secretarial applicants. Since no Negroes were then enrolled in these schools, none were hired.[48]

In Indianapolis, another major concern was recruiting heavily in 1951. A former member of the employment staff described the situation:

> In 1951, this company was hiring males at a fairly high rate of approximately 100 per week. During initial screening interviews personnel interviewer would indicate on applications "NW" for nonwhites and "W" for whites. There was little or no interest in putting Negroes on the roll so they were rarely hired.
>
> During this period it was noted that one or more of the same well-dressed Negroes were always in the lobby observing who was sent to the shop for interviews and taking obvious notes. This continued for approximately 30 days, during which time we later learned, several Negro organizations contacted our New York headquarters.
>
> In approximately another 60-90 days we were advised to hire a specific number of Negroes. I cannot remember the exact number but it was something like 25. These people were immediately hired.
>
> It would appear that this direct pressure was responsible for these people being put on the roll, and without it they most likely would not have been employed.[49]

Federal and State Commission Activities

Although the Eisenhower Administration did not support federal fair employment legislation, a President's Committee on Gov-

---

[48] Interview with former employee relations manager, September 1970.

[49] Interview, September 1970.

ernment Contracts was established on August 13, 1953 by Executive Order 10479[50] to encourage nondiscrimination in employment by government contractors. Known as the "Nixon Committee," because it was chaired by the then Vice President, this Committee developed the investigatory and fact-finding techniques later expanded by the Kennedy and Johnson Administrations as a means of increasing Negro employees among government contractors. A small sample of electrical machinery concerns, including selected plants of major companies, were among those that were investigated and prodded by the Nixon Committee and from which racial employment data were gathered.

Table 11 shows data developed by the Nixon Committee inspectors for the last three years of the decade of the 1950's. In general, these corroborate the already noted census data that show the paucity of Negroes in the industry, and the extreme shortage of Negro salaried employees. The variations from year to year are not significant except in the white collar group, but even there the increases may be partially attributable to different plants or companies sampled. The bulk of the Negro white collar employment in all these years was clerical, and found in large city areas of the North. The 1950's saw increased efforts to expand employment in these jobs, but results continued to be small.

Actions of state fair employment practice commissions also contributed to improved Negro employment in the late 1950's. In New Jersey, what was then known as the Division Against Discrimination began receiving reports of discrimination in the late 1940's and began a campaign that opened up not only an increasing number of blue collar jobs, but white collar jobs as well. The files of the New Jersey agency reveal steady improvement in this regard in successive inspection by plants of General Electric, Western Electric, McGraw-Edison, RCA, International Telephone and Telegraph, and others. Similar action occurred in Pennsylvania with Westinghouse, in New York with General Electric, in Massachusetts with Raytheon, and in Ohio with RCA and Whirlpool. In all of these states, other electrical manufacturers, including smaller ones, were also inspected and prodded toward increased hiring of Negroes.[51]

---

[50] Previous executive orders had been issued by the Roosevelt and Truman administrations, as already noted.

[51] Based upon an examination of the unpublished files of these agencies.

TABLE 11. *Electrical Manufacturing Industry*
Total and Negro Employment by Type of Employment
Various Companies and Plants, 1957, 1958, and 1959

	All Employees			White Collar Employees			Blue Collar Employees		
Year	Total	Negro	Percent Negro	Total	Negro	Percent Negro	Total	Negro	Percent Negro
1957	58,107	1,453	2.5	19,989	43	0.2	38,118	1,410	3.7
1958	70,783	1,548	2.2	30,028	224	0.7	40,755	1,324	3.2
1959	72,837	2,083	2.9	30,615	305	1.0	42,222	1,778	4.2

Source: Unpublished files of the President's Committee on Government Contracts (Nixon Committee).

Note: Data are for 4 companies, 20 plants in 1957; 5 companies, 19 plants in 1958; and 5 companies, 17 plants in 1959.

The state commissions, unlike the Nixon Committee, made few systematic company or plant analyses and none in electrical manufacturing. Nevertheless, as their number grew to include nearly all non-southern states, they helped to prepare industry for the civil rights pressures of the 1960's by opening gates for Negroes and gaining acceptance of the principles of fair employment. Several companies began a reevaluation of their racial employment policies in an effort to avoid more onerous supervision and legislation. For example, in 1954 General Electric began an internal communication and education program regarding the requirements of the Nixon Committee. Other companies similarly were moving closer toward implementing policies, particularly by employing Negro engineers, recruiting at predominantly black colleges, and reexamining recruiting methods. Affirmative action would await the 1960's. Federal and state investigations reinforced the growing business recognition that fair employment was part of good corporate citizenship, but it was not yet clear that a systematic minority manpower effort and unconventional techniques would be necessary and considered appropriate by the public.

The 1950's—Final Comment

The 1950's ended with the industry committed on a policy basis to equal opportunity, but still a long way from reaching an approximation to the goal. With only 3.3 percent of its employees black, it had come a long way from its exclusionist practices of twenty years before, adding almost 30,000 Negroes to its rolls in a decade (Table 10). Electrical manufacturing still employed far fewer blacks than many other industries and the obstacles to equal employment remained great. The older northern plants contained the bulk of black employees in 1960, some having a blue collar ratio of 10 percent or more.[52] The newer plants, however, were located in areas where Negro population was lower, where transportation from center city concentrations of Negroes was difficult, where employment agencies discriminated in placements, or where competition for jobs favored the better educated white. In addition, advancing technology insured that thousands of new jobs would be available only to those with strong educational backgrounds.

---

[52] Data in authors' possession.

Nevertheless, the situation was not without promise for Negroes. By 1960, even though the electrical machinery industry had only a small number of Negro employees who were concentrated in low skill jobs, some Negroes were employed in almost every job category. Management was generally committed, if passively, to fair employment. The industry continued to need increasing numbers of employees and many plants were located where Negro applicants could realistically be sought out. The 1950's had seen barriers broken in the salaried and supervisory jobs without any major difficulties. With its expanding employment and receptive attitude, the industry appeared ready for further progress.

CHAPTER IV

Affirmative Action in the 1960's

By 1960, it was apparent that something more than nondiscrimination would be required to increase the representation of Negroes in industrial jobs. If this was true for industry generally, it was especially the case for the burgeoning electrical machinery industry. Its large percentage of professional and salaried positions, its pleasant working conditions and good wages, and its growth gave the industry the capacity to attract an ample, highly qualified labor supply. Given the relatively disadvantaged educational status of Negroes and their inexperience in industry, they were not able to compete on even terms. This was the message of the final report of the President's Committee on Government Contracts (Nixon Committee),[53] a message immediately picked up by the incoming Kennedy Administration.

GOVERNMENTAL LEADERSHIP, THE ECONOMY, AND CORPORATE RESPONSE

The years after 1960 saw a marked change in civil rights efforts. Pressure from Negro groups for better jobs and more employment opportunity intensified. Demonstrations and riots created a new sense of urgency.

The federal government was a direct catalyst for change in minority manpower policies in the sixties. Prior to 1961, few companies maintained organized quantitative data on racial employment, or admitted so doing if they existed. (Western Electric was an exception.) In some states racial categorization of employees was forbidden. Whatever the reason, lack of such quantitative data greatly inhibited knowledge and corrective action. President Kennedy's Executive Order No. 10925 in 1961 differed most significantly from earlier presidential orders creating fair employment practices commissions by requiring all companies contracting with the federal government to submit

---

[53] See, e.g., President's Committee on Government Contracts, *Seventh Report* (Washington: Government Printing Office, 1960).

periodic compliance reports giving the number of Negroes employed and the types of jobs they filled, and by introducing the "affirmative action" concept. Most electrical machinery firms were covered by this order.

Also in 1961, the federal government asked the cooperation of business in Plans for Progress—a program for improving the employment situation of Negroes through voluntary company initiative. RCA, Western Electric, General Electric, and Westinghouse all signed soon after the program was announced, pledging to take "affirmative action" to ensure that their nondiscrimination policies were implemented. Some plans also included a promise to make a deliberate effort to recruit more Negro applicants. Although Plans for Progress has been subject to considerable criticism because performance has often not matched its public promise, the program is significant as a symbol of management consensus that equal employment opportunity should be publicly proclaimed as a company goal.

Title VII of the 1964 Civil Rights Act, effective as of July 1965, gave legislative assent and permanence to the principle of equal access to employment and extended coverage from the companies executing government contracts to all firms in interstate commerce beyond a certain minimal size. Later, directives for implementing Executive Order 10925, and subsequant orders, and findings in suits under Title VII have continued to emphasize positive recruitment of blacks and have increasingly considered an analysis of the racial composition of the work force as an important indicator of employment policy.

The More Systematic Corporate Approach

The fundamental managerial policy difference marking off the sixties from earlier years is that a significant part of industry became publicly and actively engaged in efforts to change employment practices, with the acknowledgement that new and unconventional methods were to be developed and were necessary to achieve integration within any reasonable span of time. Growing company involvement and new federal directives led gradually to a systematic evaluation of the total hiring process. Attention turned from preventing overt discrimination to eliminating the more subtle practices which might unintentionally preclude Negro applicants from qualifying for jobs. Pre-employment tests were analyzed for validity as predictors of job performance and many which measured qualities not directly necessary for the job were discarded or replaced. Personnel practices workshops stressed the importance of eliminating unconscious discrimination in employment interviews. Black interviewers were employed to "reach" Negro applicants, and a host of other tools were implemented to obtain Negro workers.

For example, companies began to question whether their traditional sources of recruitment reached potential Negro applicants, and some began to experiment with increased employment advertising in the Negro press and with conscious efforts to overcome the widespread assumption that electrical was still a "white" industry. The same approach was applied to promotions and selection for training programs. Western Electric reviewed its personnel files in 1961 to identify qualified Negroes who might have been overlooked in earlier promotions. Several other concerns inventoried their employees with a conscious effort to upgrade the existing black labor force. Efforts may have been incomplete in quality, scope, and success. Yet it was now recognized that traditional methods, usually without intent, were perpetuating or at least not altering sufficiently rapidly the racial composition of the industry.

Perhaps most important, the compliance reports themselves served as a form of internal audit by which firms were forced to assess their own experience in the matter of racial employment. They provided a tool for making judgments of results in the minority employment area as they were accustomed to do in other areas such as production and sales. Although quantified data alone do not provide sufficient information for judging the quality of a fair employment effort, they are essential in giving employment practice concreteness and rationality.

In some companies the compliance reports were used for internal evaluation. Corporate staff discussed the reports with each field location, urged realistic and rising goals, and offered staff support in solving special problems. It was to the company's advantage to retain the initiative in increasing the quantity and quality of Negro employment and thereby to reduce onerous supervision. As federal standards were raised and as executive recognition of the urgency of the problems increased, more companies developed serious internal equal employment efforts.

With emphasis on minority manpower increasing, equal employment opportunity became a more important function at the corporate level in the biggest electrical companies. In the 1960's most companies assigned responsibility for compliance reports and for Plans for Progress to a top executive, usually the vice president of personnel or industrial relations with staff support from the corporate personnel department. Western Electric and RCA created interdivisional committees of top executives to emphasize the importance of the policy and to coordinate implementation. General Electric added a major component in its relations department. Gradually specific full-time equal employment opportunity positions were developed in other companies. Western Electric, for example, created its first full-time position at the managerial level in 1963. Several years later both RCA and Western Electric set up separate departments with this responsibility at the corporate level. Westinghouse loaned an executive full time to the Plans for Progress staff in 1965-1966, and then made him head of the company's equal employment opportunity section at corporate headquarters. Other companies alloted increasing equal employment opportunity responsibility to staff and managers already assigned to personnel. Parallel developments took place in plants and field locations, with the development of full-time equal opportunity managers at most division levels and some full-time staff in the large plants.

The development of these new structures within the business organization of the firm does not in itself guarantee dramatic progress. The effectiveness and the style of such a department is determined by the priorities which the top company executives set, and by the relative weight to equal opportunity considerations in the balancing of many competitive values. The presidents of all the Big Four have made strong statements on fair employment to their management and personnel staffs, a fact which has increased the potential that this commitment will be implemented. At the very least the rise of corporate departments means that some companies are willing to commit staff positions and departmental expenses to planning and implementation in the area of equal employment opportunity.

THE OVERALL PICTURE, 1964-1969

In 1960, the United States Census reported that 3.3 percent of the electrical machinery workers were black (Table 10, Chapter III). Table 12, summarizing Appendix Tables A-1 to A-4, shows that the percentage steadily rose in the 1960's, so that by 1969 it stood at 7.3 percent. A cautionary note is required, however, because the data utilized in this chapter are not strictly comparable

TABLE 12. *Electrical Manufacturing Industry*
Percent Negro Employment by Sex and Occupational Group
United States, 1964, 1966, 1967, and 1969

	All Employees				Male				Female			
Occupational Group	1964	1966	1967	1969	1964	1966	1967	1969	1964	1966	1967	1969
Officials and managers	0.3	0.4	0.6	0.9	0.3	0.4	0.6	0.8	0.5	1.8	2.5	3.5
Professionals	0.6	0.7	0.8	1.0	0.6	0.7	0.8	1.0	1.5	2.0	1.9	1.8
Technicians	1.7	2.2	2.7	3.1	1.7	2.2	2.6	3.1	2.2	2.6	3.8	4.1
Sales workers	0.4	0.2	0.2	0.6	0.2	0.2	0.2	0.5	2.0	1.2	0.8	1.6
Office and clerical	1.8	2.0	2.4	3.9	1.8	2.3	3.0	4.4	1.7	1.8	2.1	3.7
Total white collar	1.1	1.3	1.6	2.3	0.9	1.2	1.4	1.8	1.7	1.8	2.2	3.6
Craftsmen	2.1	3.0	3.8	4.2	1.8	2.6	3.3	3.7	7.2	7.6	10.0	10.2
Operatives	6.1	8.4	9.9	11.5	5.9	8.5	9.7	11.0	6.3	8.3	10.1	11.9
Laborers	6.1	9.9	11.9	12.7	7.1	11.4	13.9	13.8	4.8	8.6	10.3	11.9
Service workers	16.2	17.5	18.5	19.2	15.9	17.8	18.7	19.4	18.1	15.7	17.0	17.8
Total blue collar	5.6	7.8	9.3	10.5	5.1	7.3	8.5	9.3	6.2	8.4	10.2	11.9
Total	3.8	5.4	6.4	7.3	3.2	4.6	5.3	5.8	5.0	6.8	8.3	9.8

Source: Appendix Tables A-1 to A-4.

to census data. The latter are collected by enumerators from individuals; the data utilized in this chapter are based on reports filed by companies with the Equal Employment Opportunity Commission or with other governmental agencies. Moreover, the 1964 data include only companies that had government contracts, while in 1966, 1967, and 1969 those with 100 or more employees are included. Nevertheless, the size of the samples for all years of the 1960's would indicate that they are quite accurate, and the upward trend of Negro employment is consistent with both our field observations and with trends in industry generally. We feel, therefore, that it is reasonable to assume that the proportion of Negro employment in electrical manufacturing did indeed more than double between 1960 and 1969.

Most of the Negro gains came in the latter part of the decade. For example, between 1966 and 1969, some 35,639 Negroes were added to the industry's rolls, an increase of 38.7 percent. (See Appendix Tables A-2 and A-4.) Obviously, such a large increase did not occur either by chance, or merely by adhering to a nondiscrimination policy. It was primarily attributable to the combination of the need for labor in a rapidly expanding economy and to the impact of stronger governmental civil rights policies. The industry now actively sought out and employed black workers as part of the affirmative action programs of various companies, some of which are discussed later in this chapter.

ANALYSIS OF OCCUPATIONAL GROUPS

Most of the increase in the number of Negroes in the industry up to 1969 came in the blue collar area. Table 12 shows that by that year only 2.3 of the white collar workers in the industry were black in contrast to the blue collar situation where 10.5 percent were Negro. Table 13 which shows the distribution of all employees and of Negroes by occupational group, points up the concentration of Negroes in the blue collar jobs, and within the blue collar group, in the lower rated jobs. Thus, whereas in 1969, 38.6 percent of all employees in the industry were employed in white collar occupations, and 61.4 in blue collar jobs, the comparative figures for Negroes were 12.1 percent and 87.9 percent. Within the blue collar area, 38 percent of all employees were operatives and 11.3 percent laborers and service workers. Among Negro workers, 60 percent were operatives and 21 percent were laborers and service workers. Thus 49.3 per-

TABLE 13. *Electrical Manufacturing Industry*
Percent Distribution of Employment by Race, Sex, and Occupational Group
United States, 1969

	All Employees		Male		Female	
Occupational Group	Total	Negro	Total	Negro	Total	Negro
Officials and managers	7.9	1.0	12.3	1.8	0.6	0.2
Professionals	10.1	1.4	15.8	2.8	0.8	0.1
Technicians	7.1	3.1	10.4	5.5	1.6	0.7
Sales workers	1.4	0.1	2.0	0.2	0.4	0.1
Office and clerical	12.1	6.5	5.8	4.3	22.5	8.5
Total white collar	38.6	12.1	46.3	14.6	25.9	9.6
Craftsmen	12.1	6.9	17.9	11.3	2.5	2.6
Operatives	38.0	60.0	26.6	50.2	56.9	69.6
Laborers	9.7	16.9	7.1	16.8	14.1	17.0
Service workers	1.6	4.1	2.1	7.1	0.6	1.2
Total blue collar	61.4	87.9	53.7	85.4	74.1	90.4
Total	100.0	100.0	100.0	100.0	100.0	100.0

Source: Computed from Table A-4.

cent of the total work force in our sample was in the lowest three classifications, but 81 percent of the black work force was so classified. The reasons for the wide discrepancies between the occupational distributions of white and Negro employees are best highlighted by an analysis of black employment in each of the occupational groups.

Officials and Managers

Electrical companies recently have been recruiting Negroes very actively for foreman training and management development programs but Negroes still held only 0.9 percent of all managerial jobs in 1969 (Table 13), and most of these were first line supervisors. Although the industry is now eager for applicants, improvement will be slow. Traditionally, executive jobs were virtually closed to Negroes in this country and training was neither sought nor supplied. In the electrical industry, recruitment is further complicated by the frequent need for managers with engineering or scientific backgrounds and the tremendous shortage of trained Negro personnel in these fields.

Traditionally also, managers have won their jobs by working up through the production, marketing, or financial departments. Production personnel are heavily engineering oriented and trained; finance men usually have an accounting background; marketing personnel are often graduates of schools of business who have worked in sales and sales management.

There is also a shortage of Negroes trained in accounting and finance. Sales is a special problem that will be discussed below. The most hopeful sign is the tremendous increase in the number of black students now enrolled in the prestigious schools of business. Whereas just a few years ago institutions like the Wharton School of Finance and Commerce of the University of Pennsylvania and the Harvard Graduate School of Business Administration had very few Negroes enrolled in their master of business administration programs, today these institutions have a 5-10 percent black student enrollment. These institutions have actively recruited Negroes, many of whom are on scholarships provided by industry or by foundations. Black graduates of MBA programs are actively recruited by electrical and other companies and are providing future candidates for high management positions, but such programs take time to achieve their results.

Electrical manufacturing companies had placed Negroes in personnel departments, particularly to handle aspects of the equal employment program. This position can also be utilized as a training job for other management opportunities and should provide a means to enter higher management or positions with broader scope.

Since the mid-1960's, a number of Negroes have been promoted from the blue collar ranks to supervisory jobs. Most of these supervise racially mixed areas. Upgrading from the ranks is another means of entering management, but again the road upward is slow because the higher the position, the more keen is the competition and the fewer are the openings. All in all, it is likely to be one or two decades at least before Negroes achieve parity in the officials and managers category, and even then only if companies persevere in affirmative action.

Professionals and Technicians

The highly scientific and engineering content of the electrical machinery industry is demonstrated by the fact that more than 17 percent of its employees are professionals or technicians (Table 13). Although electrical manufacturing concerns were among the first actively to recruit at Howard University and at other Negro engineering schools, results have often been disappointing. Less than 5 percent of the Negro employees were found in these categories in 1969.

As late as the mid-1960's, General Electric found some Negroes unconvinced that good jobs were available. Thus, a vice president of the company declared:

> One of the problems which has come up again and again in discussions with educators whose student bodies include large numbers of Negroes is the lack of recognition of the real opportunities which are open in industry—not just General Electric, but industry as a whole—for qualified Negro applicants.
>
> In the lower grades, this lack of recognition seems to be one of the factors which can breed apathy toward academic achievement. In college, it can predispose many competent young people to shy away from courses which would lead to careers in industry.
>
> "It isn't enough for us to *tell* them about good job opportunities," one of our recruiters was told last summer. "You have to *show* them." [54]

General Electric prepared a booklet describing 50 case histories of black employees and managers at all levels working in the company and distributed it to 60,000 guidance teachers and career counselors throughout the country. Such efforts are still needed to close the credibility gap.

Perhaps more important, and constantly frustrating to those concerned with affirmative action and equal opportunity, is the sad state of primary and secondary education in the inner city and southern small town or rural schools. It is here that learning is discouraged and future careers so often blighted. For many Negroes to achieve a professional career, family patterns of nonachievement must be overcome, basic skills acquired without assistance at home, and a school which defies learning must be conquered. Companies in the electrical industry have spent thousands of dollars aiding school dropouts, helping programs such as the Opportunities Industrialization Centers, and providing basic education programs to functionally illiterate prospective employees. Inevitably, however, the schools and the basic learning institutions must do a more effective job if discrimination and unequal opportunity are not to be perpetuated.

---

[54] Virgil B. Day, "Progress in Equal Employment Opportunity at General Electric," in Northrup and Rowan (eds.), *The Negro and Employment Opportunity,* p. 161.

In the technical area, it is often possible to use employees with less than a full college education. Moreover, blue collar employees can sometimes be upgraded to such positions. It is thus not surprising that with less stringent requirements there are a higher proportion of Negroes utilized as technicians than as professionals. Electrical manufacturing companies have made a special effort to upgrade Negroes to these jobs, and have had some success, although the proportion still remains small.

Nevertheless the outlook is something less than promising for substantial increases in the proportion of black professionals and technicians because of the total shortage of Negro engineers. Thus a careful analysis by the leading student of Negro opportunities in the engineering profession reported to the U.S. Department of Labor in October 1970:

> During the last 8 years there has been virtually no increase in the percentages of blacks in engineering education except for the special programs that some colleges have instituted to encourage and to retain these students.
>
> About 2 percent of all engineering students in the U.S. are black. The percentage of blacks in Bachelor of Technology programs is about 5 percent as is also the approximate percentage of blacks in technicians programs.
>
> There seems to be no question but that there are wide-spread education and employment opportunities for blacks in engineering and in technicians' work. About half of the black engineering students are in predominantly black colleges that have engineering programs which illustrates the importance of these 6 colleges. Much of the demand for black engineers and technicians has come through government fair employment regulations. This has resulted in some dissatisfaction among engineers and technicians who feel they have been hired for their color.
>
> The relative lack of information on engineering coupled with employment discrimination of the past seem to be the chief reasons for the apparent lack of interest of blacks in this profession today.[55]

One possibility to combat this shortage of engineering and technical talent is the new "cluster program," introduced by the Big Four and other major electrical concerns. This program has representatives of industry working in close cooperation with the colleges in their plant areas to develop curriculum changes, improve equipment utilization, improve budget development, etc. Encouragement of professional studies is a principal object of

---

[55] Robert Kiehl, *Opportunities for Blacks in the Profession of Engineering* (Newark: Foundation for the Advancement of Graduate Study in Engineering, 1970), p. 14.

the program. Obviously, programs of this character are essential if the availability of black professionals and technical personnel is to be increased.

Sales

A very small proportion of employees in the electrical industry —less than 2 percent—are classified as salesmen. Much of the "selling" is actually done by managers and professionals because of its technical nature. Retailing is largely in the hands of dealers who are not employees of the electrical manufacturing concerns.

Even when an employee of an electrical manufacturer is classified as sales, he probably needs a technical or engineering background in order to perform his duties. Given the few Negroes so qualified, this severely limits Negro participation in sales work. In addition, much of selling is social in nature. Until very recently, industry was too fearful of customer reaction to employ Negro sales personnel except in special situations. This is slowly changing; now all major electrical companies have Negroes on their sales forces and in their sales training programs, but the technical requirements limit the numbers and overcoming them will take time.

Office and Clerical

Office and clerical positions offer the easiest access for unskilled or poorly educated personnel to white collar positions. Here again, however, it frequently requires special efforts—affirmative action—to alter racial employment patterns. Recruitment has to be expanded and minorities have to be sold on the fact that jobs are indeed available. Obviously, offices and plants in cities having a sizeable Negro population are the best sources of office and clerical positions for black candidates.

When electrical machinery and other companies began concerted efforts to employ Negro secretaries and office help in the late 1950's and early 1960's they found that the traditional sources could supply an insufficient number of qualified candidates. Too often a black girl would apply with a diploma from a center city high school attesting that she had passed a "commercial" course and was proficient in typing and/or shorthand. Simple, objective tests would show, however, that the candidate not only was seriously below standards in both, but also needed considerable remedial work in spelling, punctuation, and grammar, and perhaps office grooming and social adaptation thereto as well. A student who was rejected after completing a high school course often believed that discrimination was the reason, whereas in fact she had been literally cheated by inadequate schooling and training.

To overcome these problems, company training has been undertaken by the major electrical companies. In New York, General Electric, Western Electric, and other companies joined with the Urban League to sponsor a training course for secretaries and clerical personnel that has been quite successful in placing personnel. In numerous other cities, companies have worked to overcome training deficiencies in order to qualify black candidates. With the high attrition typical of such jobs, there are usually openings. As a result of affirmative action, the percentage of Negroes in this occupational group is slowly but steadily increasing. Now that it is well known throughout most Negro communities that black candidates are being sought and welcomed, a substantial number apply either for training, or where qualified, for jobs. It is likely that in the 1970's the black composition of the electrical machinery companies' clerical staffs will continue to expand and will more closely reflect the racial composition of the area in which the facilities are located.

Craftsmen

The need for craftsmen in the electrical industry has been growing, as over 12 percent of the work force was in this category in 1969. Because of a shortage of qualified applicants, and because of federal pressure to include Negroes in all job classifications, many electrical plants have training programs for which Negroes are sought. This is a recent development, however. Until the 1960's Negroes were not recruited in large numbers by industry; even today, they have been effectively excluded from experience outside of industry by the discriminatory policies of many of the metal and building craft unions.

Of course some building construction unions have a higher percentage of Negro members than the percentage of craftsmen in the electrical manufacturing industry. Moreover since slavery days there has existed a tradition of craft work among Negroes particularly in the South. Unfortunately, black craftsmen are concentrated in the older crafts, particularly the trowel trades (bricklaying, plastering, and cement finishing), but also in carpentry and painting. The craft unions in these trades have a higher percentage of Negroes than do others. On the other hand, in the newer electrical, plumbing, and metal trades, Negroes for years were systematically excluded from training, from unions, and from jobs. And it is these later trades and skills, not the older ones, which the electrical manufacturing industry primarily requires.

As in other industries, the electrical industry has traditionally drawn its craftsmen from three sources: company apprentice programs; employment of those who learned their skills elsewhere, including immigrants; and upgrading or on-the-job training of those with production experience. The apprentice programs traditionally attracted "a high type person—the high school graduate . . . who did not make it to college because of family attitudes or finances." [56] Such persons today are more likely to go to college, a fact which has contributed substantially to the shortage of apprentice candidates. Moreover, Negroes similarly situated are also likely to choose the socially more prestigious college training to that of apprenticeship.

For black candidates willing to go into apprentice training, the inequities of inner city or segregated southern secondary schools remain serious problems. Deficiencies in mathematics, especially, hurt their opportunities to gain acceptance or to progress rapidly if they are admitted to programs. Moreover, far fewer Negroes have parents or relatives who are craftsmen, especially in the metal or electrical trades, than do whites. Not having been raised in a tradition of craft work during their formative years limits their pre-training knowledge of the work and of the opportunities that are available. Moreover, given past practice, Negroes generally have assumed that craft work was not available to them and have not applied or trained for it. Even though companies are now seeking them, it is difficult to reverse past practice, thinking, and modes.

A better source of Negro applicants is found among employees. They have a knowledge of the job and its opportunities and know the rewards that await the trained mechanic. Companies have instituted special training programs to encourage black employees to register for apprentice training. Although the proportion of Negroes applying for such training, even with

---

[56] Northrup, Rowan, *et al., Negro Employment in Basic Industry,* Part Two, p. 81. This analysis closely follows that of Dr. Northrup's in the similar situation involving automobile industry apprentice requirements.

special company encouragement, often remains below that of whites in relation to their relative numbers in the work force, there has been a notable rise in Negro apprentices and craftsmen since 1964. The total percentage is still, of course, relatively small.

Besides having to overcome deficiencies in education and experience, the Negro production worker who enters craft training may well encounter hostility from his white counterparts, some of whom resent the "intrusion" into what was once a white stronghold. He is likely to feel more compatible remaining in the production jobs where fellow black employees are found in much larger numbers.

Moreover, production pay is high and craft training often does not result in higher earnings immediately. As Dr. Northrup has noted:

> This demands both a long-run view, strong motivation, and confidence in the future. Such apprentice candidates do not sacrifice their seniority, but apparently this is not enough motivation. Discrimination and motivation go hand in hand. Because of the pervasiveness of the former in the life of Negroes, the latter is often lacking. Opportunities for training which require sacrifice for eventual improvement are not grasped. Inexperience in industry, lack of family help in setting goals, the absence of similar opportunities in previous generations, and the difficulty of breaking with the past all continue to reduce the number of Negroes who grasp the opportunities for skilled trades training.[57]

Since few Negroes are found in the key mechanical and electrical trade unions and are virtually nonexistant among immigrants who have craft training, existing employees are likely to remain the largest source of future craftsmen. Assuming continued company affirmative action designed to increase the number and proportion of black craftsmen, the future is likely to see a slow, but certain increase in Negroes trained for these key jobs. Perhaps as the numbers increase, their demonstrated success will induce more to enter the trades and speed the movement.

Operatives

Operative jobs cover a wide variety of functions, some very low skilled, others requiring considerable training. Most of the jobs can be mastered with a very little to a moderate amount

[57] *Ibid.*, Part Two, p. 84.

of training. Most involve assembly work or light machine tending. By far the largest number of electrical machinery employees —38 percent—were classified as operatives in 1969. Moreover, 50.2 percent of all Negro males and 69.6 percent of all Negro females were so classified in that year (Table 13).

Obviously Negroes have encountered little difficulty in obtaining operative jobs in recent years. These are entry positions, but are necessary for gaining experience for better jobs in the future. The large numbers of Negroes employed in these jobs and the continuing recruitment of black employees by the electrical machinery companies not only insures a rising percentage of Negroes in the industry, but also provides the potential for upgrading to line supervision, craftsman positions, and such white collar jobs as technician.

Laborers and Service Workers

Only 11.3 percent of all employees in the industry were classified as laborers or service workers in 1969. There is some upgrading from these positions, but many are deadend jobs and therefore appropriate only for people with limited abilities or education.

Negroes are overrepresented in these jobs, as in industry generally. In many areas it is difficult to employ whites, despite attempts at integration, since the jobs are known as "Negro work." Because of the small numbers involved, however, the problem does not seem significant for the future.

Female Employment

As noted in Chapter II, females in the electrical industry are primarily concentrated in two job categories—office and clerical, where in 1969 22.5 percent of all females were found, and operatives where 56.9 percent worked. A third category, laborers, employed 14.1 percent of all females (Table 13).

Black females were likewise concentrated in these three classifications, but the distribution was different—8.5 percent as office and clerical, 69.6 percent as operatives, and 17 percent as laborers. Yet despite this skewed distribution and despite the fact that there were more black men than black women in the industry in 1969, the percent of Negro females in the industry was substantially greater than that of males—9.8 percent as compared with 5.8 percent (Table 12). The reasons for this are found in the nature of jobs performed by men and women. Relatively few women—only about 3 percent of all women employees—are found in the top three white collar categories, while 25.1 percent of all male employees occupy these positions. These are also the jobs which require the greatest training and experience required, and in which, for reasons already noted, the fewest Negroes are to be found.

On the other hand, Negro females are in great demand for clerical work, and for small assembly jobs in which women have been found to be very adept. Partly because of this factor and the difficulties of Negro men in obtaining craft training noted above, the proportion of Negro women in the industry is higher than that of Negro males for both blue collar and white collar employment.

Intraplant Mobility

Movement up the job hierarchy in the electrical industry is relatively fluid. There are few long, narrow seniority lines, except in some plants which make heavy utility apparatus. The large proportion of semiskilled operative jobs provide opportunities for the unskilled and inexperienced to work at entry jobs and to advance as they gain knowledge and experience. Seniority systems allow broad movement within large work areas. Few if any "black departments" or segregated work areas have ever existed, and we have no record of seniority agreements especially designed to confine Negroes to specific jobs, as was common in such industries as tobacco or paper manufacturing. The situation is remarkably similar to that described by Dr. Northrup in the aerospace industry:

> In general, the wider the seniority district, the greater the opportunitiy for Negroes for upgrading and advancement. As the most recently hired and the group which has the fewest skills, Negroes profit substantially from broad opportunities for movement. But, of course, a wide seniority progression system has its corollary disadvantage, too. When employment turns downward, it provides the broadest opportunities for bumping. With Negroes both relatively new and still overly concentrated in the semiskilled and relatively unskilled jobs, they are then especially vulnerable.

Job bidding is widely used in the industry to fill higher jobs. Company officials have repeatedly complained to the author that Negroes appear more reluctant to bid on jobs than do whites, and that Negroes often require great encouragement if they are to bid. The situation is similar to the already noted disproportionate reluctance of Negroes to take advantage of training opportunities. Again the reasons appear to be lack of experience in industrial practice, fear of moving from a job situation which is acceptable to white fellow workers to one which is not, or lack of motivation.[58]

Most supervisors in the electrical machinery industry are promoted from the ranks. Companies have sought out Negroes for such promotions, and as more become qualified, the movement into these jobs by black production workers should increase.

LOCATIONAL FACTORS

We have noted that locational factors have traditionally acted to reduce Negro participation in the industry. Once Negroes migrated northward during World War II and thereafter, location became less of a bar. The tendency to build the new electronic facilities in suburban "industrial parks" or in other outlying locations accessible only by automobile has again been detrimental to Negro employment, but there are also offsetting factors as neighborhoods change. Thus "Nela Park," headquarters of General Electric's lamp division in East Cleveland, Ohio, was once far from that city's black neighborhoods. Today, Nela Park is in a black neighborhood.

Where Negroes are concentrated in the inner city and plants are in the subsurbs, Negro employment tends to lag in most instances.[59] Thus the plants on Route 128 near Boston or in Valley Forge near Philadelphia have much more difficulty recruiting Negroes than do plants, sometimes of the same company, which are located within an easy bus or subway ride of areas where Negro dwellings are concentrated. Black office and clerical personnel, in particular, have no difficulty in finding jobs with banks, insurance companies, and business offices in the cities and therefore do not need to travel to the suburbs.

Many electrical machinery plants are still located in smaller cities and towns of the North, Midwest, and West which have few black inhabitants. Others are found outside of the city limits of these towns, particularly the new ones in the South, Midwest, and West. Public transportation is not usually available

[58] *Ibid.*, Part Three, p. 184.

[59] The drug manufacturing industry has been able to attract Negroes to suburban plants by reason of its unusually good conditions and special locational factors. See F. Marion Fletcher, *The Negro in the Drug Manufacturing Industry*, in the Racial Policies of American Industry series, pp. 76-82.

to such locations, a fact which can preclude some of the disadvantaged from gaining employment. Again, comments made in discussing the situation in the aerospace industry are appropriate here:

> Negroes who do find a home in the suburbs or outer county areas, or who are willing to commute long distances, are usually both well qualified for their jobs and highly motivated. They are, therefore, capable both of finding work and of gaining promotion and upgrading opportunities.
>
> Nevertheless, despite some exceptions, the farther from center city is the aerospace plant, the fewer Negroes are found on its employment rolls. Workers who have had little experience, motivation, or assistance do not know enough to look for work at a great distance. The expense of long commutes cuts deep into the income from low-skilled and entry jobs, and the frustrations of getting to work that is far from home easily discourage the inexperienced or uninitiated commuter. Moreover, superior workers can often find work closer to home. Consequently, both employment and upward job movement in the industry are restricted by the necessities of plant location, and its typical distance from the central city where Negroes are more and more concentrated.[60]

REGIONAL INFLUENCES

We noted that one reason why the electrical industry was slow to employ Negroes was that it had its roots in New England and smaller cities of the Middle Atlantic region, where few Negroes dwelled prior to World War II. Since then, however, we have witnessed not only the migration of the electrical and other industries into all regions of the country, but also massive migrations of black people from the South to the North, Midwest, and West Coast. The various regions of America are themselves changing enormously. Job openings have encouraged thousands of Negroes to migrate to areas where once only whites dwelled. A careful study, by regions, of minority manpower data is essential for understanding the total Negro employment picture. Especially it will give us clues as to where black employment is likely to increase in the future and where specific efforts will be required to bring this about.

New England

Once the center of the electrical industry, New England is now far outshadowed by the Middle Atlantic and the Midwest. As an area where the Negro population is small, it is to be ex-

---

[60] Northrup, Rowan, *et al., Negro Employment in Basic Industry,* Part Three, p. 188.

pected that black electrical machinery employment would be correspondingly low in New England. Such is the case, with Negro employment well below the national average. (Compare Tables 12 and 14.) Insofar as the occupational distribution is concerned, Table 15 shows that in 1969 the overall situation in New England is slightly better than that nationally, approximately 85 percent of the Negroes were in blue collar jobs as compared with only about 60 percent of all employees.

Minority employment efforts in New England are greatly hampered by the concentration of Negroes in the central cities of the largest metropolitan areas, and the location of electrical plants in small towns and on the outskirts of large cities. Over one-half the electrical employees in the region are in the greater Boston metropolitan area, which is about 10 percent black. The bulk of the Negroes, however, live in the central city, while electrical manufacturing plants are located in small towns 20 to 30 miles away, particularly along Route 128. Few Negroes live in these outlying areas which are not easily accessible by public transportation. A number of such plants have recently been established in the small communities of southern New Hampshire and Vermont, where the population is almost all white. An increasing number of Negroes do dwell in the southern tier of Connecticut, and this is reflected in increasing black employment in such cities as Bridgeport.

Middle Atlantic

The Middle Atlantic region—New York, New Jersey and Pennsylvania—has long been a center of the electrical manufacturing industry. Employment in plants in these states has continued to expand, with Negroes obtaining an increasing proportion of the jobs. By 1969 over 7 percent of all employment in the area was black (Table 16). The data in Table 17 also show that the distribution of Negro employees within plants in these states is more favorable than that nationally and in most regions.

There are several reasons why this is so. Many of the older electrical manufacturing plants are found in areas of substantial Negro population. Thus Western Electric has major manufacturing facilities in the Newark, New Jersey area; General Electric in Newark and Trenton, New Jersey, and Philadelphia; Westinghouse in the Newark area, Philadelphia, and Pittsburgh; RCA in the Newark area and in Camden, New Jersey. Smaller

TABLE 14. *Electrical Manufacturing Industry*
Percent Negro Employment by Sex and Occupational Group
New England Region, 1964, 1966, 1967, and 1969

	All Employees				Male				Female			
Occupational Group	1964	1966	1967	1969	1964	1966	1967	1969	1964	1966	1967	1969
Officials and managers	0.2	0.2	0.5	0.6	0.1	0.2	0.4	0.5	2.1	1.3	2.2	2.8
Professionals	0.6	0.5	0.6	0.8	0.6	0.5	0.6	0.8	1.2	—	0.8	0.8
Technicians	1.2	1.6	1.7	2.7	1.2	1.6	1.7	2.5	0.8	1.3	1.8	4.2
Sales workers	0.7	0.2	0.2	0.5	0.2	0.1	0.1	0.4	4.7	0.7	0.6	0.7
Office and clerical	0.8	1.0	1.3	2.1	1.0	1.1	1.4	2.4	0.7	0.9	1.2	2.0
Total white collar	0.7	0.8	1.0	1.5	0.7	0.8	0.9	1.3	0.8	0.9	1.2	2.1
Craftsmen	1.9	2.4	3.0	3.6	1.2	1.7	2.1	2.8	11.0	13.2	15.8	11.4
Operatives	3.4	4.1	5.8	5.7	2.8	4.2	5.4	5.9	3.8	4.1	6.1	5.7
Laborers	5.5	8.3	7.6	6.4	5.6	7.8	11.1	9.5	5.5	8.5	6.4	5.5
Service workers	7.0	5.5	7.2	6.8	6.9	5.3	6.9	6.9	7.4	7.1	9.9	5.9
Total blue collar	3.5	4.4	5.6	5.5	2.6	3.7	4.7	5.2	4.3	5.1	6.4	5.8
Total	2.4	3.1	3.9	3.9	1.6	2.3	2.8	3.1	3.5	4.2	5.2	4.9

Source: Appendix Tables A-5 to A-8.

Note: For regional definition, see Table 8.

TABLE 15. *Electrical Manufacturing Industry*
Percent Distribution of Employment by Race, Sex, and Occupational Group
341 Establishments, New England Region, 1969

Occupational Group	All Employees		Male		Female	
	Total	Negro	Total	Negro	Total	Negro
Officials and managers	7.6	1.2	13.3	2.3	0.5	0.3
Professionals	11.5	2.5	20.0	5.3	0.8	0.1
Technicians	6.9	4.7	11.2	9.2	1.4	1.2
Sales workers	1.7	0.2	2.4	0.3	0.7	0.1
Office and clerical	12.3	6.7	6.2	4.8	19.9	8.3
Total white collar	40.0	15.3	53.1	21.9	23.3	10.0
Craftsmen	10.0	9.1	16.2	14.8	2.0	4.6
Operatives	39.2	57.7	24.2	45.6	58.4	67.3
Laborers	9.1	14.9	3.9	11.8	15.7	17.4
Service workers	1.7	3.0	2.6	5.9	0.6	0.7
Total blue collar	60.0	84.7	46.9	78.1	76.7	90.0
Total	100.0	100.0	100.0	100.0	100.0	100.0

Source: Computed from Appendix Table A-8.

concentrations of Negroes, but nonetheless substantial ones, are also found in Buffalo and Syracuse, New York, where Westinghouse and General Electric, respectively, have plants. In addition, many smaller electrical manufacturing concerns are found in Middle Atlantic cities which have sizeable black populations.

The electrical manufacturing industry also has a number of key plants in areas of New York and Pennsylvania where few Negroes dwell. The overall regional percentages combine plants with some of the highest and some of the lowest percentages of Negroes in the industry. The former, however, are among the largest plants in the industry. Given the population mix in the region, therefore, one would expect a continued increase in Negro employment.

Another reason why this region leads and has a favorable employment ratio is that many of the plants are older ones which opened their doors to Negro employment for labor market and wartime policy reasons during World War II. As Negro migration to the plant areas continued, an increase in black employ-

TABLE 16. *Electrical Manufacturing Industry*
Percent Negro Employment by Sex and Occupational Group
Middle Atlantic Region, 1964, 1966, 1967, and 1969

Occupational Group	All Employees				Male				Female			
	1964	1966	1967	1969	1964	1966	1967	1969	1964	1966	1967	1969
Officials and managers	0.4	0.6	0.8	1.1	0.4	0.6	0.7	1.0	0.3	3.4	5.1	5.8
Professionals	0.8	0.8	1.0	1.2	0.7	0.8	1.0	1.2	2.1	1.1	2.4	2.2
Technicians	2.1	2.8	3.7	4.1	2.1	2.8	3.3	4.0	2.4	3.4	7.5	5.0
Sales workers	0.6	0.4	0.5	0.6	0.3	0.2	0.3	0.4	3.1	2.0	1.8	2.1
Office and clerical	2.2	2.7	3.2	4.8	2.3	2.9	3.6	4.8	2.1	2.6	3.0	4.7
Total white collar	1.4	1.8	2.2	2.8	1.2	1.5	1.8	2.2	2.1	2.7	3.3	4.6
Craftsmen	2.6	3.3	4.4	5.0	2.4	3.2	3.8	4.4	9.9	6.7	15.3	16.6
Operatives	7.9	10.9	11.9	12.1	6.0	9.2	9.9	9.8	9.8	12.3	13.6	14.4
Laborers	7.0	13.3	12.4	12.3	7.1	13.7	12.7	11.7	7.0	13.0	12.1	12.8
Service workers	11.5	14.2	14.5	16.4	11.4	14.9	14.9	16.5	11.8	10.8	12.0	15.5
Total blue collar	6.6	9.5	10.4	10.5	5.0	7.6	8.2	8.2	9.3	12.3	13.4	14.2
Total	4.3	6.3	7.2	7.3	3.1	4.7	5.3	5.4	6.9	9.5	10.6	11.2

Source: Appendix Tables A-9 to A-12.

Note: For regional definition, see Table 8.

TABLE 17. *Electrical Manufacturing Industry*
Percent Distribution of Employment by Race, Sex, and Occupational Group
847 Establishments, Middle Atlantic Region, 1969

	All Employees		Male		Female	
Occupational Group	Total	Negro	Total	Negro	Total	Negro
Officials and managers	8.7	1.3	12.6	2.3	0.6	0.3
Professionals	10.7	1.8	15.3	3.4	0.9	0.2
Technicians	7.6	4.3	10.4	7.7	1.8	0.8
Sales workers	1.4	0.1	1.9	0.1	0.5	0.1
Office and clerical	13.2	8.6	6.3	5.6	27.7	11.7
Total white collar	41.6	16.1	46.5	19.1	31.5	13.1
Craftsmen	14.7	10.0	20.7	16.7	2.2	3.3
Operatives	33.5	55.6	24.4	44.1	52.2	67.2
Laborers	8.4	14.3	6.1	13.1	13.4	15.4
Service workers	1.8	4.0	2.3	7.0	0.7	1.0
Total blue collar	58.4	83.9	53.5	80.9	68.5	86.9
Total	100.0	100.0	100.0	100.0	100.0	100.0

Source: Computed from Appendix Table A-12.

ment followed plant expansion. The fact that these three states were among the first to adopt fair employment practice laws also aided and encouraged integration. Finally, several companies in these areas—including all of the Big Four—pioneered with outreach recruiting and training programs.

Tables 16 and 17 also show that the Middle Atlantic region has a more favorable distribution of Negro employment than most regions. To be sure, Negroes are concentrated in the lower rated jobs as usual, and they are overrepresented especially in the laborer and service worker categories. Although in each of our sample years the Middle Atlantic region increased its percentage of Negroes in each occupational group, the most rapid increases were in the less skilled jobs. Nevertheless, the Middle Atlantic region also has the highest black participation rate in all white collar categories except professionals, where it closely trails the West Coast; and it is well above the national average in the percent of Negroes in skilled and semiskilled blue collar jobs. Here again the early movement of the plants toward integration and the long seniority of some of the black employees are factors in the more favorable occupational distributions.

An interesting aspect of electrical machinery employment in the Middle Atlantic region is the above industry average proportion of white collar employment. This reflects in part the large number of corporate headquarters in the area, and also the fact that major companies have their research facilities there. In part, this reduces the overall proportion of Negroes because they are relatively less available for professional and managerial jobs. On the other hand, electrical manufacturing companies, as well as other concerns, have their central EEO components at their headquarters, and this presence, plus the desire to show progress in integration, usually increases opportunities for Negroes in these offices. It is likely that the better than average representation in the white collar, as well as in the craftsman category, probably also reflects the more effective recruiting of the Big Four.

As in other regions, Negro females are better represented than Negro males in the Middle Atlantic electrical machinery plants, but the disparity is greater. One reason is the already noted concentration of headquarters and the large numbers of Negro female clerical employees which are now found in these offices. Another is that employment expansion has been much greater in recent years in the plants producing products requiring light electronic assembly than in those making heavier products. The former utilize a large proportion of women, and this has resulted in a substantial increase in the proportion of black females.

Midwest

The Midwest includes areas such as Cleveland and Chicago, where large numbers of Negroes dwell, and also the Great Plains states in which the percentage of black population is very low. The older electrical manufacturing plants are found in the cities where the Negro population is concentrated, but many newer plants are not. With some exceptions, equal employment opportunity began later in the Midwest than in the Middle Atlantic states. The more conservative bent of this area is reflected by the fact that the states in the Midwest, as a whole, adopted fair employment practice laws a decade or more after those in the Middle Atlantic.

The electrical manufacturing industry is flourishing in the Midwest, where the greatest number of employees work. Its percentage of Negroes lagged behind that of the Middle Atlantic region until 1969, when it rose to 7.5 percent (Table 18). Negro job distributions are lower than the national average for all white collar categories except that of officials and managers where they are equal to the low national average (Table 19). Blue collar distributions are similar to the national averages. The percentages of Negroes in the various occupational categories improved in each of the sample years.

The number and percentage of black employees in plants in the Midwest vary greatly with the location. Chicago's black population in 1967 was estimated at 30 percent,[61] but Negroes comprised only 14 percent of electrical workers.[62] In that year, however, 53 percent of all Negroes working for the industry in the Midwest were in Chicago, compared to 26 percent of all employees. In Chicago, black participation in the semiskilled and unskilled jobs was particularly high, but it was also high for craftsmen (11.1 percent) and above average for officials and managers and technicians in the white collar category.[63] Major facilities having large percentages of Negroes are Western Electric's Hawthorne works and the Chicago operations of General Electric manufacturing Hotpoint appliances.

Outside of Chicago, high black participation is found in plants in Indianapolis, St. Louis, Detroit, Columbus, and Cleveland—all cities with sizeable Negro populations. Cleveland is the home of General Electric's lamp division, Western Electric has a major facility in Indianapolis, Westinghouse in Columbus, and Emerson Electric, General Electric, Century and others have plants in St. Louis. In smaller towns in the Midwest and in the Minneapolis-St. Paul area, where Honeywell has its headquarters and largest employee concentration, the Negro population and the Negro plant labor force are much lower.

A better grasp of the racial employment situation in the Midwest can be gained by comparing the situation in its subregions.

---

61 U.S. Bureau of Labor Statistics, *The Social and Economic Status of Negroes in the United States, 1969,* Report 375, 1970, p. 9.

62 U.S. Equal Employment Opportunity Commission, *Job Patterns for Minorities and Women in Private Industry, 1967,* Report No. 2 (Washington: The Commission, 1970), Vol. 2.

63 *Ibid.*

TABLE 18. *Electrical Manufacturing Industry*
Percent Negro Employment by Sex and Occupational Group
Midwest Region, 1964, 1966, 1967, and 1969

	All Employees				Male				Female			
Occupational Group	1964	1966	1967	1969	1964	1966	1967	1969	1964	1966	1967	1969
Officials and managers	0.2	0.4	0.7	1.1	0.2	0.3	0.7	1.0	0.2	1.3	2.1	4.1
Professionals	0.3	0.5	0.8	1.1	0.3	0.4	0.7	1.0	0.5	5.1	2.5	3.1
Technicians	1.0	1.9	2.5	2.8	1.0	1.9	2.4	2.7	1.2	2.4	3.5	4.2
Sales workers	0.1	0.2	0.1	0.9	0.1	0.1	0.1	0.8	0.3	1.0	0.5	2.4
Office and clerical	1.7	1.7	2.2	4.4	1.6	1.7	2.1	4.0	1.7	1.8	2.2	4.5
Total white collar	0.8	1.1	1.5	2.5	0.6	0.9	1.3	1.8	1.6	1.8	2.2	4.4
Craftsmen	1.9	3.0	4.1	4.4	1.5	2.5	3.2	3.6	8.1	11.9	13.4	14.0
Operatives	6.0	8.7	9.5	10.3	6.2	9.0	9.5	10.5	5.9	8.4	9.5	10.2
Laborers	5.2	9.3	11.6	11.5	6.3	10.8	14.6	13.6	3.3	8.1	9.3	9.6
Service workers	16.1	16.2	16.1	15.7	15.3	16.4	16.1	16.0	21.6	15.4	15.9	14.4
Total blue collar	5.4	8.1	9.2	9.7	5.2	7.8	8.8	9.2	5.7	8.5	9.6	10.2
Total	4.0	6.2	7.0	7.5	3.6	5.5	6.2	6.5	4.8	7.2	8.3	9.0

Source: Appendix Tables A-14 to A-16.

Note: For regional definition, see Table 8.

TABLE 19. *Electrical Manufacturing Industry Percent Distribution of Employment by Race, Sex, and Occupational Group 1,003 Establishments, Midwest Region, 1969*

	All Employees		Male		Female	
Occupational Group	Total	Negro	Total	Negro	Total	Negro
Officials and managers	7.1	1.0	11.5	1.7	0.5	0.2
Professionals	6.1	0.9	9.8	1.5	0.5	0.2
Technicians	5.1	1.9	7.8	3.2	0.9	0.4
Sales workers	1.5	0.2	2.4	0.3	0.3	0.1
Office and clerical	10.7	6.2	5.2	3.2	19.0	9.6
Total white collar	30.5	10.2	36.7	9.9	21.2	10.5
Craftsmen	11.6	6.7	17.8	9.9	2.1	3.3
Operatives	43.4	59.9	33.3	53.8	58.7	66.5
Laborers	12.9	19.8	10.1	21.2	17.2	18.4
Service workers	1.6	3.4	2.1	5.2	0.8	1.3
Total blue collar	69.5	89.8	63.3	90.1	78.8	89.5
Total	100.0	100.0	100.0	100.0	100.0	100.0

Source: Computed from Appendix Table A-16.

Table 20 shows percent employment by race and occupational distribution in the eastern and western sectors of this region for 1969. The eastern sector, which includes the great Midwest industrial states of Illinois, Indiana, Michigan, Ohio, and Wisconsin, contains the major black population centers of the area. Not surprisingly, the black ratio in these states is 8.1 percent as compared with 4.8 percent in the western region, where few Negroes live. More than 80 percent of the industry's Midwest employment is, however, in states that have a substantial black population—a fact which augurs well for the future of Negro employment in the area.

Qualitatively, however, both Midwest subregions rank below the Middle Atlantic area and the national average, although the more populous eastern sector's situation is superior to the western one. Again, the slower adoption of fair employment practice by plant managements and of fair employment laws by states are undoubtedly causal factors for this lag.

West Coast

The West Coast electrical machinery industry is heavily related to aerospace. Many of the plants located there were founded to supply electronic gear to missle and aircraft plants. The significance of this sector of the industry is indicated by the fact that 13 percent of the employment in the industry was concentrated in the three Pacific Coast states in 1969. This is the area, of course, where the largest segment of the aerospace industry is concentrated.

Like the aerospace industry, the electrical machinery facilities on the West Coast are highly oriented toward research and development. Consequently, the percentage of white collar work in these plants is well above average. White collar employment comprises more than 56 percent of all work there as compared with less than 40 percent of all jobs nationally. (See Table 8, Chapter II.)

Negroes comprised only 4.1 percent of the West Coast work force in 1969, yet nearly 25 percent of those employees—twice the ratio of the national average—held white collar jobs (Tables 21 and 22). Their lack of overwhelming percentages in the lower rated blue collar jobs and the tremendous percentage of white collar work combined to give Negroes an underrepresentation in spite of their relatively greater participation in the white collar jobs. Despite some success in recruiting Negro technical and professional personnel, West Coast electrical concerns, like their aerospace counterparts, have found it difficult to continue to find qualified candidates, even when successful recruiting is done relatively early.

Electrical employment on the West Coast is very concentrated and again the concentration of black employees is even greater. Almost 90 percent of all Negroes employed on the West Coast in 1967 worked in the Los Angeles area as compared with about one-half of all employees.[64] This pattern probably reflects population distributions, but also the preference of many persons, black or white, for southern California living.

As a matter of fact, electrical manufacturing employment on the West Coast closely reflects the aerospace industry in terms of racial patterns as well as of occupational patterns. Both have been quite successful in recruiting the limited number of Negroes available for professional jobs, and both began their recruit-

[64] *Ibid.*

TABLE 20. *Electrical Manufacturing Industry Employment by Race, Occupational Group, and Subregion Midwest Region, 1969*

	East North Central[a] (815 establishments)			West North Central[b] (188 establishments)		
Occupational Group	Total	Negro	Percent Negro	Total	Negro	Percent Negro
Officials and managers	33,043	406	1.2	7,539	33	0.4
Professionals	26,946	288	1.1	7,809	78	1.0
Technicians	22,142	684	3.1	6,845	120	1.8
Sales workers	7,626	80	1.0	1,090	1	0.1
Office and clerical	50,262	2,368	4.7	11,105	311	2.8
Total white collar	140,019	3,826	2.7	34,388	543	1.6
Craftsmen	53,686	2,521	4.7	12,348	354	2.9
Operatives	202,260	22,345	11.0	46,119	3,333	7.2
Laborers	64,539	8,077	12.5	9,410	417	4.4
Service workers	7,738	1,151	14.9	1,459	291	19.9
Total blue collar	328,223	34,094	10.4	69,336	4,395	6.3
Total	468,242	37,920	8.1	103,724	4,938	4.8

Source: Preliminary Equal Employment Opportunity Commission data, 1969.

[a] Illinois, Indiana, Michigan, Ohio, Wisconsin.

[b] Iowa, Kansas, Minnesota, Missouri, Nebraska, South Dakota.

TABLE 21. *Electrical Manufacturing Industry*
Percent Negro Employment by Sex and Occupational Group
West Coast Region, 1964, 1966, 1967, and 1969

Occupational Group	All Employees				Male				Female			
	1964	1966	1967	1969	1964	1966	1967	1969	1964	1966	1967	1969
Officials and managers	0.4	0.5	0.6	0.9	0.4	0.5	0.5	0.8	—	1.0	1.2	1.5
Professionals	0.8	0.9	1.0	1.1	0.8	0.9	1.0	1.1	1.0	1.0	1.3	1.1
Technicians	2.5	2.4	2.6	2.9	2.4	2.5	2.7	2.9	4.1	1.7	1.6	2.0
Sales workers	0.3	0.3	0.1	0.4	—	0.2	0.1	0.3	2.3	0.6	—	1.2
Office and clerical	1.3	1.8	2.2	2.6	2.3	3.3	4.0	4.7	0.9	1.3	1.6	1.9
Total white collar	1.3	1.4	1.6	1.8	1.3	1.4	1.6	1.8	1.2	1.3	1.5	1.8
Craftsmen	2.0	3.4	3.1	3.0	1.9	3.1	3.0	2.6	3.0	5.4	3.9	5.0
Operatives	4.8	8.0	8.6	7.8	3.9	7.7	9.9	8.6	5.3	8.1	7.6	7.3
Laborers	3.7	9.5	10.5	9.3	5.6	12.2	10.9	10.5	1.4	6.5	10.2	8.5
Service workers	22.5	24.2	24.2	21.6	23.4	23.7	24.1	22.1	14.3	29.4	25.3	17.1
Total blue collar	4.4	7.2	7.7	7.0	4.1	6.7	7.7	6.8	4.8	7.9	7.7	7.3
Total	2.6	3.9	4.2	4.1	2.2	3.3	3.7	3.6	3.4	5.1	5.1	5.1

Source: Appendix Tables A-17 to A-20.

Note: For regional definition, see Table 8.

TABLE 22. *Electrical Manufacturing Industry Percent Distribution of Employment by Race, Sex, and Occupational Group 548 Establishments, West Coast Region, 1969*

Occupational Group	All Employees		Male		Female	
	Total	Negro	Total	Negro	Total	Negro
Officials and managers	9.9	2.1	14.5	3.4	0.7	0.2
Professionals	18.7	5.3	27.0	8.6	2.0	0.4
Technicians	11.2	7.8	15.1	12.4	3.4	1.4
Sales workers	1.5	0.1	2.0	0.2	0.5	0.1
Office and clerical	15.1	9.6	5.7	7.5	34.1	12.7
Total white collar	56.4	24.9	64.3	32.1	40.7	14.8
Craftsmen	12.2	9.0	15.4	11.2	5.9	5.8
Operatives	24.9	47.8	15.6	37.7	43.4	62.1
Laborers	5.3	12.2	3.1	9.3	9.7	16.2
Service workers	1.2	6.1	1.6	9.7	0.3	1.1
Total blue collar	43.6	75.1	35.7	67.9	59.3	85.2
Total	100.0	100.0	100.0	100.0	100.0	100.0

Source: Computed from Appendix Table A-20.

ment early.[65] At the present, both industries are suffering severe cutbacks, probably with considerable adverse effects on their black workers.

South

The South is developing as a major electrical manufacturing center and shows great promise of becoming one of the leading areas of the industry. By 1969, 19.4 percent of all electrical manufacturing employees were found in the South (Table 8). Moreover, southern plants in that year had the largest percentage of Negroes of any region (Table 23).

Southern electrical manufacturing plants are heavily oriented toward light electronic assembly and tend to be concentrated in the southeastern states—the South Atlantic subregion. Thus, they are in areas where large numbers of Negroes dwell and

---

65 See Northrup, Rowan, *et al.*, *Negro Employment in Basic Industry,* Part Three, pp. 199-204, for the southern California aerospace situation.

provide many jobs which can be mastered with little training. As noted earlier, the southern sector of the industry began after World War II, and initially employed a largely white labor force. This occurred primarily because the South was a labor surplus area and because most managers gave little thought to employing Negroes when whites could easily be hired. Employment standards were set high and since Negroes were disproportionately less well educated and experienced, they found great difficulty in competing for the available jobs even when available to them.

Beginning especially in 1966, major electrical manufacturing firms in the South began their affirmative action programs. At about the same period, the labor market tightened substantially. The results are shown in Table 23. The ratio of employment rose from 4.2 percent to 11.3 percent between 1964 and 1969. Of special significance was the increase in black female employment from 4.6 to 15.6 percent at this time.

In the South Negro female employment has traditionally lagged in factory employment, except in the tobacco industry.[66] Moreover, most southern electrical manufacturing plants are located outside of cities and towns away from Negro housing concentrations, and are impossible to reach except by automobile. This has added an obstacle to Negro employment. Prior to 1967, the percentage of Negro females in the southern sector of electrical manufacturing was below the national average. By 1967, it had surged ahead, indicating the significance of changing employer policies and the tightening labor market.

The increases between 1967 and 1969 indicate that the improvement in the employment ratio of Negroes in the southern electrical manufacturing industry will continue, and perhaps at a faster pace. Professor Richard L. Rowan's companion study of the textile industry details the sudden change of the industry in the mid-1960's from one which virtually excluded Negroes (and, as the largest southern manufacturing industry, had set the standards for southern industry) to an industry that is becoming a major—perhaps eventually the largest—employer of blacks.[67] Already, thousands of southern whites who received their first factory experience in the textile industry are employed

---

66 See Northrup, Rowan, *et al.*, *Negro Employment in Southern Industry*, Part Three, for a discussion of the racial employment pattern in tobacco.

67 *Ibid.*, Part Five.

TABLE 23. *Electrical Manufacturing Industry*
Percent Negro Employment by Sex and Occupational Group
South Region, 1964, 1966, 1967, and 1969

	All Employees				Male				Female			
Occupational Group	1964	1966	1967	1969	1964	1966	1967	1969	1964	1966	1967	1969
Officials and managers	0.1	0.2	0.3	0.6	0.1	0.2	0.3	0.6	—	0.8	0.8	1.9
Professionals	0.3	0.6	0.6	0.8	0.3	0.5	0.5	0.8	2.9	5.4	2.9	1.6
Technicians	1.2	2.0	2.4	3.2	1.1	1.8	2.3	2.9	1.9	4.3	3.5	5.7
Sales workers	0.3	0.2	0.3	0.5	0.3	0.1	0.3	0.4	1.0	2.3	—	1.4
Office and clerical	2.2	2.0	2.5	4.2	1.6	2.5	3.5	5.3	2.4	1.8	2.0	3.6
Total white collar	1.0	1.2	1.5	2.3	0.6	1.0	1.3	1.8	2.4	2.0	2.1	3.7
Craftsmen	1.2	2.2	3.6	4.2	1.3	2.1	3.5	4.1	1.0	2.9	4.2	5.4
Operatives	6.1	7.9	12.2	17.8	7.5	8.9	12.2	16.4	5.3	7.1	12.2	18.8
Laborers	7.3	9.6	14.7	19.7	9.6	11.7	15.6	18.3	4.2	6.8	13.7	21.0
Service workers	32.8	31.3	34.7	36.8	31.3	31.0	34.0	35.7	59.6	34.5	42.0	46.1
Total blue collar	6.2	7.8	11.7	16.2	7.1	8.4	11.1	13.7	5.2	7.1	12.3	18.9
Total	4.2	5.6	8.2	11.3	3.9	5.4	6.7	8.4	4.6	6.0	10.3	15.6

Source: Appendix Tables A-21 to A-24.

Note: For regional definition, see Table 8.

in newer, better paying industries, including electrical manufacturing. It is very likely that this industry is now providing a similar training for Negroes and that many black employees, including, as was the case with whites, large numbers of women, will seek jobs in industries such as electrical manufacturing to better their incomes. The jump from 10.3 to 15.6 percent Negro female employment in two years would seem to bear this out.

Qualitatively, however, the South continues to lag. Table 24 shows that in 1969 only 7.2 percent of its black employees were in salaried positions, substantially below the national average. It is apparent that the electrical manufacturing industry has, like industry generally, concentrated first on overcoming its prior failure to employ many Negroes in its southern plants. The more difficult job of improving the occupational mix was barely begun by 1967 and showed only slight improvement by 1969, although electrical manufacturing was well ahead of such traditionally southern industries as tobacco, lumber, textiles, or

TABLE 24. *Electrical Manufacturing Industry*
Percent Distribution of Employment by Race, Sex, and Occupational Group
693 Establishments, South Region, 1969

	All Employees		Male		Female	
Occupational Group	Total	Negro	Total	Negro	Total	Negro
Officials and managers	6.9	0.4	11.2	0.8	0.3	*
Professionals	10.1	0.8	16.3	1.6	0.7	0.1
Technicians	7.0	2.0	10.4	3.6	1.9	0.7
Sales workers	0.9	*	1.2	0.1	0.3	*
Office and clerical	10.8	4.0	5.7	3.6	18.6	4.3
Total white collar	35.7	7.2	44.8	9.7	21.8	5.1
Craftsmen	10.7	4.0	16.5	8.0	2.0	0.7
Operatives	42.1	66.5	28.4	55.4	62.8	75.6
Laborers	10.1	17.6	8.2	17.8	13.0	17.5
Service workers	1.4	4.7	2.1	9.1	0.4	1.1
Total blue collar	64.3	92.8	55.2	90.3	78.2	94.9
Total	100.0	100.0	100.0	100.0	100.0	100.0

Source: Computed from Appendix Table A-24.

* Less than 0.05 percent.

paper, none of which had even 3 percent of their black employees in salaried positions in 1968.[68] Given the propensity of the better educated southern Negroes to look to the North for jobs and those in the North to remain there, finding qualified personnel to expand black salaried employment participation is not easily accomplished. Nevertheless, the goals of equal employment opportunity require improvement and this is likely to occur, but slowly.

As was noted, the bulk of southern electrical manufacturing employment is concentrated in the South Atlantic states. Most of that, in turn, is found in the northern tier of this area, although Florida has considerable such employment. Insofar as Negro employment participation is concerned, however, there was little difference in 1969 from one southern region to another. Table 25 shows that the overall black ratio is nearly identical for the Southeast and East South Central subregions, and slightly greater for the smallest area in terms of employment, the West South Central (Southwest) subregion. In the Southeast, however, Negroes have a greater share of salaried jobs, although that percentage remained small in 1969. Overall, one would expect that the burgeoning electrical manufacturing industry in the Southeast will steadily improve its utilization of Negroes, quantitatively and qualitatively, in part because of civil rights pressures, in part because of tightening labor markets in an area of industrial expansion.

IMPACT OF COMPANY SIZE

Tables 26 and 27 compare Negro employment in the four largest firms—General Electric, RCA, Western Electric, and Westinghouse—with that in the overall industry for 1967 and 1969. Of special interest is the fact that the Big Four lagged behind the industry in the proportion of blacks in 1967, but forged ahead two years later, not only in the overall figure, but in each occupational category except that of craftsmen as well. This occurred despite relative occupational structures which would lead one to believe that the Big Four would actually lag in relation to the industry as a whole.[69]

[68] *Ibid.*, Part Six.

[69] The Big Four and overall industry data are less than strictly comparable because the former include non-electrical manufacturing plants, but it is not believed that this results in any significant distortion.

TABLE 25. *Electrical Manufacturing Industry*
Employment by Race, Occupational Group, and Subregion
South Region, 1969

Occupational Group	South Atlantic[a] 369 Establishments			East South Central[b] 162 Establishments			West South Central[c] 162 Establishments		
	Total	Negro	Percent Negro	Total	Negro	Percent Negro	Total	Negro	Percent Negro
Officials and managers	12,273	79	0.6	5,337	29	0.5	5,831	34	0.6
Professionals	18,782	197	1.0	4,309	26	0.6	11,110	67	0.6
Technicians	12,705	507	4.0	2,506	59	2.4	8,548	202	2.4
Sales workers	1,558	9	0.6	658	3	0.5	712	3	0.4
Office and clerical	19,994	1,005	5.0	6,194	130	2.1	10,615	398	3.7
Total white collar	65,312	1,797	2.8	19,004	247	1.3	36,816	704	1.9
Craftsmen	19,760	779	3.9	10,771	455	4.2	5,855	284	4.9
Operatives	61,020	10,619	17.4	43,291	6,143	14.2	38,393	8,641	22.5
Laborers	11,794	3,135	26.6	19,072	3,050	16.0	3,351	550	16.4
Service workers	2,054	790	38.5	1,784	413	23.2	1,052	597	56.7
Total blue collar	94,628	15,323	16.2	74,918	10,061	13.4	48,651	10,072	20.7
Total	159,940	17,120	10.7	93,922	10,308	11.0	85,467	10,776	12.6

Source: Preliminary Equal Employment Opportunity Commission data, 1969.

[a] Delaware, District of Columbia, Florida, Georgia, Maryland, North Carolina, South Carolina, Virginia, West Virginia (Southeast).

[b] Alabama, Kentucky, Mississippi, Tennessee.

[c] Arkansas, Louisiana, Oklahoma, Texas (Southwest).

TABLE 26. *Electrical Manufacturing Industry Employment by Race and Occupational Group Big Four and Total Industry, 1967*

	Big Four Companies[a]			Total Industry		
Occupational Group	Total	Negro	Percent Negro	Total	Negro	Percent Negro
Officials and managers	58,590	330	0.6	134,885	842	0.6
Professionals	82,715	782	0.9	195,542	1,640	0.8
Technicians	50,563	1,223	2.4	132,836	3,568	2.7
Sales workers	10,676	41	0.4	23,682	55	0.2
Office and clerical	101,084	3,309	3.3	229,935	5,515	2.4
Total white collar	303,628	5,685	1.9	716,880	11,620	1.6
Craftsmen	105,363	2,553	2.4	221,942	8,422	3.8
Operatives	261,778	29,249	11.2	792,630	78,758	9.9
Laborers	39,029	4,999	12.8	172,411	20,499	11.9
Service workers	9,252	1,591	17.2	27,997	5,167	18.5
Total blue collar	415,422	38,392	9.2	1,214,980	112,846	9.3
Total	719,050	44,077	6.1	1,931,860	124,466	6.4

Source: Data in authors' possession and Appendix Table A-3.

[a] General Electric, Western Electric, Westinghouse, and RCA.

TABLE 27. *Electrical Manufacturing Industry Employment by Race and Occupational Group Big Four and Total Industry, 1969*

Occupational Group	Big Four Companies[a]			Total Industry		
	Total	Negro	Percent Negro	Total	Negro	Percent Negro
Officials and managers	65,940	666	1.0	137,731	1,265	0.9
Professionals	97,173	1,521	1.6	176,851	1,844	1.0
Technicians	54,189	2,214	4.1	124,664	3,921	3.1
Sales workers	14,805	227	1.5	24,438	158	0.6
Office and clerical	108,696	6,579	6.1	211,348	8,243	3.9
Total white collar	340,803	11,207	3.3	675,032	15,431	2.3
Craftsmen	116,756	4,750	4.1	210,937	8,861	4.2
Operatives	259,687	37,246	14.3	664,749	76,593	11.5
Laborers	38,532	6,886	17.9	169,545	21,616	12.7
Service workers	8,662	1,752	20.2	27,411	5,253	19.2
Total blue collar	423,637	50,634	12.0	1,072,642	112,323	10.5
Total	764,440	61,841	8.1	1,747,674	127,754	7.3

Source: Data in authors' possession and Appendix Table A-4.

[a] General Electric, Western Electric, Westinghouse, and RCA.

The Big Four employ a greater proportion of their workers in salaried jobs, particularly at the professional and managerial levels for which, as we have noted, disproportionately fewer Negroes are qualified. Their larger than average Negro white collar ratio is based primarily on their employment of substantial numbers of black clerical personnel. The Big Four companies in addition lead in each of the salaried occupational groups. As a matter of fact, the great bulk of the black managerial, professional, and technical employees in the industry work for the Big Four or for other large firms. Extensive Big Four recruiting and affirmative action have more than offset the greater need for scarce black personnel in positions in these categories and have, as a result, enabled these concerns to pace the industry.

The data in Tables 26 and 27 also reveal that the Big Four did not employ a higher percentage of black craftsmen than did the industry generally. The craftsman ratio may be a function of the product mix. The Big Four do most of the utility apparatus and very advanced electronic work, some of which requires a higher proportion of craft work, in relation to total employment, and again the traditional shortage of Negroes in this occupation is reflected in the percentage of blacks.

The operative group, like that of office and clerical, again finds the Big Four taking the lead. These, together with the bottom two classifications, are the positions where affirmative action results show up the most quickly and where the demands of the tight labor market have helped to expand black employment most rapidly. Affirmative action results are obviously more difficult to achieve in the higher rated jobs where training and experience mean so much.

Within the Big Four, there is some variation. Two companies are very close to the industry average and two are well above. The differences were probably strongly influenced by locational factors. A larger percentage of employees in two of the companies are within the labor markets of heavily black populated areas. This accessibility of potential black employees to certain large plants appears to have made a significant difference in the overall percentage of Negroes in these companies. A similar locational concentration of companies in other industries was found to be determinative of which concern has the largest Negro complement.[70]

Twenty Large Firms

Table 28 includes data for twenty of the largest electrical manufacturing firms for 1968. Unfortunately, these twenty do not include all of the largest fifteen firms as we have listed them in Table 3, but they are representative of the major firms in the industry. At first glance, there seems to be a positive correlation between size and Negro employment. The range of black participation rates *within* each of the clusters, however, is rather extensive. For the four largest firms, employment of Negroes ranges from 4.9 to 8.7 percent; for the second four, from 2.2 to 18.6 percent; and for the remaining twelve, from 0.7 to 7.6 percent.[71] In fact, if one ranked the firms by black participation rate, the Big Four rank 2nd, 3rd, 7th, and 12th. The next four rank 1st, 14th, 18th, and 19th. It seems that it is the extremely high percentage of Negroes in one firm in the second group that makes the record of the second cluster of companies appear better than that of the remaining twelve. Location within a heavily Negro populated urban labor market undoubtedly is the key factor affecting that company's black employee ratio.

The range of Negro employment within the Big Four, however, is much smaller: two hold high ranks in employing blacks and only one is below the middle, and not far below. Size for this cluster does appear related to a somewhat better record in Negro employment. We find higher black participation rates in every category except "craftsmen," "officials and managers," and "laborers." Overall black participation is 0.4 percent higher than in the remaining sixteen companies. Insofar as the Big Four are concerned, two of the companies have consistently been several percentage points above it and none have been more than 1.6 percentage points behind the industry or large company average.[72]

---

[70] For example, by reason of its concentration of plants in the Detroit inner city area with its heavy black population, Chrysler has a higher percentage of Negro employment than do Ford or General Motors. See Northrup, Rowan, *et al., Negro Employment in Basic Industry,* Part Two, pp. 96-99.

[71] Data in authors' possession.

[72] Data in authors' possession.

TABLE 28. *Electrical Manufacturing Industry*
Employment by Race, Occupational Group, and Company Size
Twenty Large Companies, 1968

	Twenty Large Companies			Big Four[a]			Next Largest Four in Sample			Remaining Twelve Companies in Sample		
Occupational Group	Total	Negro	Percent Negro	Total	Negro	Percent Negro	Total	Negro	Percent Negro	Total	Negro	Percent Negro
Officials and managers	83,178	513	0.6	59,255	378	0.6	9,925	85	0.9	13,998	50	0.4
Professionals	122,331	1,227	1.0	89,462	992	1.1	18,498	160	0.9	14,371	75	0.5
Technicians	78,538	1,991	2.5	54,138	1,409	2.6	11,298	283	2.5	13,102	299	2.3
Sales workers	17,467	110	0.6	12,147	97	0.8	2,091	10	0.5	3,229	3	0.1
Office and clerical	140,991	4,312	3.1	101,936	3,735	3.7	17,504	264	1.5	21,551	313	1.5
Total white collar	442,505	8,153	1.8	316,938	6,611	2.1	59,316	802	1.4	66,251	740	1.1
Craftsmen	142,713	4,284	3.0	108,708	3,113	2.9	12,738	542	4.3	21,267	629	3.0
Operatives	358,184	36,627	10.2	243,208	28,178	11.6	53,224	3,897	7.3	61,752	4,552	7.4
Laborers	73,973	10,257	13.9	42,229	6,595	15.6	11,662	2,258	19.4	20,082	1,404	7.0
Service workers	15,694	2,535	16.2	9,407	1,888	20.1	3,365	282	8.4	2,922	365	12.5
Total blue collar	590,564	53,703	9.1	403,552	39,774	9.9	80,989	6,979	8.6	106,023	6,950	6.6
Total	1,033,069	61,856	6.0	720,490	46,385	6.4	140,305	7,781	5.5	172,274	7,690	4.5

Source: Data in authors' possession.

[a] General Electric, Western Electric, Westinghouse, and RCA.

We may therefore conclude from our comparison by size that although the very largest companies may not all lead in employment of Negroes, they will not fall very far below the industry average. One important factor is that they are all large enough to have plants in very diverse labor markets. The enormous range (0.7 to 18.6 percent) in other major, but much smaller companies, in contrast, is partly explained by the labor market areas within which the firms operate: a firm operating in a semirural area with a virtual absence of Negroes in the employment pool has the lowest percentage. The highest percentage is found in a firm operating in a large metropolitan area which has a large percentage of Negroes in the labor force. Another factor affecting the Big Four is the greater attention they receive from the federal government and civil rights agencies because they are so visible. Furthermore, they also have the structure and organization to implement policy whereas a smaller firm may lack these assets.

COMPARISON WITH OTHER INDUSTRIES

Comparing electrical manufacturing with other industries is one way to evaluate its progress in racial integration. Three such comparisons are made: with all other industry, with two comparable broad industry groupings, and with six "basic" industries.

All Industry

Table 29 compares Negro employment in the electrical manufacturing industry with all industries and with all manufacturing industry reporting to the Equal Employment Opportunity Commission for 1969. It is at once apparent that the electrical industry's Negro proportion is below that both of all industry and what is more comparable, all manufacturing industry. The overall proportion of Negroes for all industry and for manufacturing industry is almost identical, but manufacturing industries have one-half the proportion of Negroes in white collar jobs. Nonmanufacturing includes small service businesses and banks and insurance companies in which Negroes have made significant strides since 1966.[73] Locational advantages also accrue to

[73] See Armand J. Thieblot, Jr., and Linda Pickthorne Fletcher, *Negro Employment in Finance*, Studies of Negro Employment, Vol. II (Philadelphia: Industrial Research Unit, Wharton School of Finance and Commerce, University of Pennsylvania, 1970).

TABLE 29. *Electrical Manufacturing Industry Compared with All Industries and Manufacturing Industries Employment by Race and Occupational Group United States, 1969*

Occupational Group	Electrical Manufacturing			All Industries Reporting to EEOC			Manufacturing Industries Reporting to EEOC		
	Total	Negro	Percent Negro	Total	Negro	Percent Negro	Total	Negro	Percent Negro
Officials and managers	137,731	1,265	0.9	2,545,601	38,090	1.5	1,129,192	11,935	1.1
Professionals	176,851	1,844	1.0	2,338,821	49,652	2.1	851,417	8,852	1.0
Technicians	124,664	3,921	3.1	1,240,602	69,432	5.6	561,496	15,523	2.8
Sales workers	24,438	158	0.6	2,472,974	98,868	4.0	394,060	6,529	1.7
Office and clerical	211,348	8,243	3.9	4,934,257	299,860	6.1	1,512,326	54,724	3.6
Total white collar	675,032	15,431	2.3	13,532,255	555,902	4.1	4,448,491	97,563	2.2
Craftsmen	210,937	8,861	4.2	3,886,454	194,636	5.0	2,403,703	126,347	5.3
Operatives	664,749	76,593	11.5	6,713,650	902,533	13.4	5,150,395	695,720	13.5
Laborers	169,545	21,616	12.7	2,605,050	566,846	21.8	1,746,021	341,500	19.6
Service workers	27,411	5,253	19.2	1,861,304	500,586	26.9	278,872	63,492	22.8
Total blue collar	1,072,642	112,323	10.5	15,066,458	2,164,601	14.4	9,578,991	1,227,059	12.8
Total	1,747,674	127,754	7.3	28,598,713	2,720,503	9.5	14,027,482	1,324,622	9.4

Source: Preliminary Equal Employment Opportunity Commission data, 1969.

nonmanufacturing positions in comparison with those in manufacturing since so many of these jobs are located in the larger cities.

Total manufacturing is a more comparable standard by which to judge Negro employment in electrical manufacturing. Again the industry falls short of the overall national average. A closer look at Table 29 shows, however, that the electrical industry is actually slightly ahead of all manufacturing in white collar employment. The deficiency is in blue collar work and results largely from the fact that a much larger percentage of laborers, and especially Negro laborers, is found in manufacturing generally than in the electrical manufacturing industry, although the latter is also below the all manufacturing standard in the craftsman and operative classifications.

Table 30 amplifies this point by comparing the occupational distribution by race for all manufacturing and electrical manufacturing in 1969. Whereas 31.7 percent of total employment and 7.4 percent of Negro employment in all manufacturing are in white collar classifications, the proportions for electrical manufacturing are 38.6 percent and 12.1 percent, respectively. On the other hand, the proportion of laborers in all manufacturing is 12.5 percent for total employment and 25.8 percent for Negro employment in comparison with 9.7 percent and 16.9 percent for electrical manufacturing. Again this points up the fact that the higher black ratio in all manufacturing is largely attributable to the higher proportion of unskilled work in other industries as compared with electrical manufacturing and the greater concentration of Negroes in these industries.

SIC 35 and SIC 37

Two broad industrial classifications, non-electrical machinery (SIC 35) and transportation equipment (SIC 37) are often compared with electrical manufacturing. There are some valid reasons for this because they employ about the same number of people as electrical, are engaged in operations of physical assembly, and are also included in the durable industries group. All three groups were employment growth industries during the mid- and late 1960's. Both non-electrical machinery and transportation equipment are, in fact, conglomerations of disparate industries placed together for statistical niceties. The combination of industries requiring such different production processes

TABLE 30. *Electrical Manufacturing Industry and All Manufacturing*
Percent Distribution of Employment by Race
United States, 1969

Occupational Group	Electrical Manufacturing		All Manufacturing	
	Total	Negro	Total	Negro
Officials and managers	7.9	1.0	8.0	0.9
Professionals	10.1	1.4	6.1	0.7
Technicians	7.1	3.1	4.0	1.2
Sales workers	1.4	0.1	2.8	0.5
Office and clerical	12.1	6.5	10.8	4.1
Total white collar	38.6	12.1	31.7	7.4
Craftsmen	12.1	6.9	17.1	9.5
Operatives	38.0	60.0	36.7	52.5
Laborers	9.7	16.9	12.5	25.8
Service workers	1.6	4.1	2.0	4.8
Total blue collar	61.4	87.9	68.3	92.6
Total	100.0	100.0	100.0	100.0

Source: Computed from Table 29.

as automobiles, aerospace, shipbuilding, and railroad car building into something called transportation equipment is more contrived than practical. Nevertheless, Table 31 compares the 1969 racial employment data for non-electrical machinery and transportation equipment with electrical manufacturing.

Insofar as the employment of Negroes is concerned, the transportation equipment industry appears to be well in advance of electrical manufacturing and non-electrical machinery. In fact, the relatively high percentage of Negroes in transportation equipment reflects the high black employment in automobile manufacturing. But transportation equipment black employees are concentrated in the blue collar areas, and especially in the semi-skilled and laborer categories. For 1969, the electrical industry had the largest numbers of Negroes in white collar occupations as a whole. Black employment in the electrical machinery industry was probably increasing a little faster than in non-

TABLE 31. *Electrical Manufacturing, Non-Electrical Machinery, and Transportation Equipment Employment by Race and Occupational Group United States, 1969*

Occupational Group	Electrical Manufacturing SIC 36			Non-Electrical Machinery SIC 35			Transportation Equipment SIC 37		
	Total	Negro	Percent Negro	Total	Negro	Percent Negro	Total	Negro	Percent Negro
Officials and managers	137,731	1,265	0.9	117,512	580	0.5	139,639	1,900	1.4
Professionals	176,851	1,844	1.0	96,159	1,043	1.1	168,270	1,398	0.8
Technicians	124,664	3,921	3.1	87,588	1,413	1.6	75,262	1,560	2.1
Sales workers	24,438	158	0.6	34,691	98	0.3	8,261	15	0.2
Office and clerical	211,348	8,243	3.9	175,388	4,516	2.6	172,441	6,634	3.8
Total white collar	675,032	15,432	2.3	511,338	7,650	1.5	563,873	11,507	2.0
Craftsmen	210,937	8,861	4.2	297,859	10,296	3.5	368,381	22,792	6.2
Operatives	664,749	76,593	11.5	448,951	40,315	9.0	615,742	113,280	18.4
Laborers	169,545	21,616	12.7	110,503	16,462	14.9	88,709	18,313	20.6
Service workers	27,411	5,253	19.2	25,533	3,458	13.5	33,476	8,322	24.9
Total blue collar	1,072,642	112,323	10.5	882,846	70,531	8.0	1,106,308	162,707	14.7
Total	1,747,674	127,754	7.3	1,394,184	78,181	5.6	1,670,181	174,214	10.4

Source: Preliminary Equal Employment Opportunity Commission data, 1969.

TABLE 32. *Seven Manufacturing Industries*
Percent Negro Employment by Industry and Occupational Group
United States, 1966 and 1968

Occupational Group	Automobiles		Aerospace		Steel		Rubber Tires		Petroleum		Chemical[a]		Electrical[b]	
	1966	1968	1966	1968	1966	1968	1966	1968	1966	1968	1966	1968	1966	1968
Officials and managers	1.2	1.4	0.4	0.6	1.0	1.3	0.6	0.9	0.2	0.3	0.2	0.4	0.4	0.8
Professionals	0.6	0.7	0.8	0.9	0.2	0.5	0.3	0.8	0.6	0.7	0.5	0.8	0.7	1.0
Technicians	1.2	1.5	1.9	2.1	1.4	2.2	1.3	2.1	1.3	1.7	1.5	2.5	2.2	2.9
Sales workers	c	c	0.3	0.6	0.3	0.3	7.5[d]	5.0[d]	2.4	2.7	0.3	0.7	0.2	0.6
Office and clerical	3.8	4.4	2.8	3.4	2.3	2.9	1.4	1.9	3.9	4.8	1.8	3.2	2.0	3.2
Total white collar	2.1	2.4	1.5	1.7	1.4	1.9	1.4	1.6	2.0	2.4	0.9	1.6	1.3	2.0
Craftsmen	3.0	3.3	4.6	5.4	5.9	6.3	2.3	2.8	1.9	2.2	2.2	2.9	3.0	4.0
Operatives	20.2	21.3	11.9	13.5	17.8	18.7	11.6	12.8	6.8	7.3	7.6	9.9	8.4	10.2
Laborers	27.6	29.1	22.4	20.7	27.7	28.7	17.1	22.8	19.7	18.0	25.8	22.4	9.9	13.0
Service workers	27.2	27.3	22.2	23.2	19.1	20.0	30.4	29.2	24.2	21.5	22.9	23.0	17.5	18.9
Total blue collar	17.6	18.6	9.0	10.0	16.1	16.7	10.8	11.7	6.3	6.2	8.0	9.2	7.8	10.0
Total	13.6	14.5	4.8	5.4	12.8	13.3	8.1	8.3	3.7	3.9	5.0	6.0	5.4	7.0

Source: Herbert R. Northrup, Richard L. Rowan, *et al.*, *Negro Employment in Basic Industry*, Studies of Negro Employment, Vol. I (Philadelphia: Industrial Research Unit, Wharton School of Finance and Commerce, University of Pennsylvania, 1970), Part Eight, p. 727; and data in authors' possession.

[a] There is a possible 1-2 point downward bias in this sample.

[b] Interpolated from 1967 and 1969 Equal Employment Opportunity Commission data.

[c] Few sales workers combined with office and clerical.

[d] Including many non-sales personnel; actually very few Negroes in this classification.

TABLE 33. *Electrical Manufacturing and Chemical Industries Employment by Race and Occupational Group United States, 1969*

Occupational Group	Electrical Manufacturing SIC 36			Chemical and Allied Products SIC 28		
	Total	Negro	Percent Negro	Total	Negro	Percent Negro
Officials and managers	137,731	1,265	0.9	110,177	934	0.8
Professionals	176,851	1,844	1.0	91,802	1,233	1.3
Technicians	124,664	3,921	3.1	58,319	2,395	4.1
Sales workers	24,438	158	0.6	49,142	489	1.0
Office and clerical	211,348	8,243	3.9	123,395	4,928	4.0
Total white collar	675,032	15,431	2.3	432,835	9,979	2.3
Craftsmen	210,937	8,861	4.2	141,657	6,614	4.7
Operatives	664,749	76,593	11.5	252,102	32,770	13.0
Laborers	169,545	21,616	12.7	80,490	18,283	22.7
Service workers	27,411	5,253	19.2	22,539	5,294	23.5
Total blue collar	1,072,642	112,323	10.5	496,788	62,961	12.7
Total	1,747,674	127,754	7.3	929,623	72,940	7.8

Source: Preliminary Equal Employment Opportunity Commission data, 1969.

electrical machinery or transportation equipment since over 14 percent of all new employees in the industry between 1966 and 1967 were black. The differences in black participation in these groups appear to be partly a function of the character of jobs—for example, the large percentage of semiskilled work in automobile manufacturing. Our next comparison will bring this out more clearly.

Six Basic Industries

Table 32 shows employment by race in the electrical industry and six other industries in 1966 and 1968. The proportion of Negroes in electrical manufacturing is, as predicted in Chapter II, slightly in excess of that in chemicals. The large percentage of Negroes in the bottom two classifications of chemicals in our 1968 sample does not quite offset the better showing of electrical manufacturing in overall white collar, and in the craftsmen and operatives jobs.

Our chemical industry data for 1968, however, do not include plants with less than 500 employees. Using 1969 EEOC data for both industries, we find that the chemical industry had forged slightly ahead of electrical by expanding its white collar representation of blacks so as to equal that of the electrical industry. This combined with the higher blue collar percentages of Negroes in chemicals, especially in the lower rated jobs, raises the 1969 overall Negro participation to a slightly higher figure than in electrical manufacturing (Table 33).

Nevertheless, electrical machinery comes fairly close in proportion of Negro employees to what one can predict on the basis of occupational distributions. How companies, unions, and government are affecting this situation is discussed in the following chapters.

CHAPTER V

The Impact of Management, Unions, and Government

The electrical manufacturing industry is encountering difficulty in further integrating its work force after the first great push added considerably to the black complement in the semiskilled operative and, to a lesser extent, the office and clerical groups. The high percentage of salaried jobs and the general occupational mix put a premium on skill and experience despite the large number of clerical and semiskilled operative jobs. Since many of the latter favor female employment, the problem of improving black male employee participation is a real one. Qualitative progress has been made, but the differential between white and black occupational distributions remains substantial.

In Chapter II we anticipated these difficulties and raised the question whether, in the years ahead, the industry's management could overcome these difficulties as it has overcome many technical and marketing obstacles to its continued growth and prosperity. As we noted, it is as yet too early to give a definitive answer to this question. It is both pertinent and of interest to note what the Big Four are doing to improve their minority employment picture and to assist in urban problems.

MANAGEMENT ORGANIZATION AND PRACTICE

It is now very clear to companies that they cannot leave racial policies to the judgment of decentralized managers. To be sure, local management can be given wide latitude for developing programs *within the framework of company policy,* and indeed should be given that latitude if programs geared to local needs are to be developed. These programs, however, must be monitored by the top executives through staff representatives, first to ascertain that they are or can be made effective, and second, to be sure that they are within the framework of company policy. In large companies, racial policies, like finance, marketing, and labor relations policies, require clear directives and a central staff both to advise on their construction and to follow up their implementation.

All the Big Four companies have recognized these facts and have developed careful policies at each location which are guided from the top. Their objective is clear: increase the quality and quantity of minority employees. This is the objective of affirmative action, and it means repeatedly in fact, giving preference to minorities when at all possible. It also means taking an active role in improving the environment so that minorities can be brought into the system. Thus all companies have participated in National Alliance of Businessmen and prior government programs to train the disadvantaged. They also are involved in programs in their key location cities to improve education and housing and numerous other activities such as the sponsorship of Opportunities Industrialization Centers. General Electric and Philco-Ford were among the earliest sponsors of these centers, but now all the Big Four and many others support OIC's.

The need for central direction and the difficulties that are caused by the failure to give careful top executive leadership is well illustrated by problems which arose in plants in which integration had made little progress. We noted in Chapter III that decentralized management guided General Electric racial policies in the South during the 1950's. The result was a few lily-white plants and others integrated for the period. This was true in the North as well as in the South. The following case history illustrates some of the problems involved in integration after national company policy took over the issue and a strong national policy was established with a central staff to guide and enforce it was organized.

Two Southern Plants

The electrical manufacturing industry has not had as serious integration problems as have some industries. There have been no discriminatory seniority systems, no "black departments" to disperse, no segregated local unions to merge. Its occupational structure permits a broad-based intraplant movement that has permitted it to avoid problems of such industries as paper or tobacco.[74] This does not, of course, mean that everything has always gone smoothly, but the required adjustments have not been as difficult as in some industries.

For example, it is commonplace knowledge that integration is smoother when it is initiated from the beginning rather than fundamentally changing existing employment conditions and relationships. In one southern plant, General Electric integrated from the start; by the time that affirmative action became an industrial byword, this plant was fully integrated and able to employ Negro white collar personnel and upgrade black workers to supervisory jobs without any problems. In another plant, however, the work force remained all white until 1961. The first black employee, a female, was subjected to considerable harassment, possibly including some by an unsympathetic supervisor. Management, however, took a firm stand and continued with its integration progress. Yet even though overt opposition dies in such an instance and blacks and whites usually get along well now, a racial transition of this character is difficult to accomplish rapidly. This is true not only in the South; it is true also in the North, especially where the new employees are not only black, but often disadvantaged as well.

Employing the Urban Disadvantaged

Difficulties of a different nature were encountered at General Electric's Chicago operations.[75] The plant there is located on the Chicago/Cicero border. Cicero is a white working or lower middle class community with a long record of hostility to any potential black residents. Traditionally, the Chicago division employed persons from this area many of whom were immigrants from middle, southern, or eastern Europe, or the second or third generation of these ethnic groups. Labor market and civil rights pressures in the 1960's, however, caused the company to add a substantial number of Negroes and to increase

---

[74] See Northrup, Rowan, *et al., Negro Employment in Southern Industry*, Parts One and Three.

[75] Based on the field study by Theodore V. Purcell and Gerald F. Cavanagh, "Alternative Routes to Employing the Disadvantaged Within the Enterprise," *Proceedings of the Twenty-Second Annual Meeting*, Industrial Relations Research Association, New York, December 29-30, 1969, pp. 66-77. See also Richard D. James, "Learning to Work: GE Finds the Unskilled Pose Problems at Start, But It Sights Progress," *Wall Street Journal*, May 20, 1968.

substantially its complement of Spanish speaking workers in its labor force at a period when many of its older employees were reaching retirement age. Many of the new workers lived on "the other side" of the company's plants in the Chicago slum areas of Garfield Park and North Lawndale.

By 1968 its labor force of 3,600, which included only 60 Negroes before the buildup, was one-third black and the Puerto Rican, Cuban, and Mexican-American complement had risen from 400 to 700. The buildup necessitated a revision of no longer applicable hiring standards. The high school education prerequisite was eliminated, standards regarding previous arrests were modified to prevent youthful or minor indiscretions from disqualifying applicants, and other measures designed to take in personnel instead of excluding them were built into the employment process.

Among those hired were many who would now be termed "disadvantaged." General Electric encountered the now familiar problems which accompany employing workers with little education, and no experience or even familiarity with industrial work. (Neither the authors nor the GE management believe that these problems are necessarily racial. Absenteeism and turnover, for example, are more prevalent among young people of any race, and most of these hired at General Electric's Chicago operations are young.)

Absenteeism at the Chicago works rose from 2 percent in 1963 to 4-8 percent in 1968. Annual turnover in the hourly work force also increased, from 18 percent in 1963 to 36 percent in 1968. Of the 1600 hires in 1968, 51 percent of the blacks and 63 percent of the whites left before the end of the year. Both turnover and absenteeism were even higher in 1966 when one of the plants ran a temporary second shift manned almost entirely by recently hired disadvantaged workers. Such problems had their inevitable reflection in cost. Rework and excessive paper work generated by turnover and other personnel problems rose drastically.

During this transition period GE experimented with different ways of dealing with the new work force. A small counseling program began in 1967 and lasted about one and a half years, achieving some reduction of absenteeism. A full-day orientation program was tried in one plant with some success but for most new employees orientation remained sketchy. Two and a half hours on the employees own time were allotted to bringing groups of 25 to 30 people through a maze of unfamiliar data about benefits, attendance policies, wage assignments, insurance and so forth. General Electric instituted courses in credit to help employees avoid wage garnishments, and also a high school equivalency program, "Hotpoint High," with the assistance of the Chicago Board of Education. (Hotpoint is the former name of this division.)

By 1969 a survey showed that most foremen thought that blacks on the average did as well in quality and quantity of work but 56 percent still saw absenteeism and turnover as a major problem. After extensive interviews with management, foremen, and a random sample of hourly employees, we found 5 major reasons for turnover:

1. Inadequate relationships with supervisors.
2. The distrust, anger, and restlessness of youth.
3. Frustration with assembly-type work as monotonous, uncreative, and dead-end.
4. Availability (during the 1960's) of other jobs in the Chicago area or attraction of easier money through welfare or even ghetto crime.
5. False expectations about wages, or failure to understand the incentives.

Some of these problems are obviously beyond the control of management and are rooted in the knotty problems of urban society. But others can be met and solved by new, creative management.

Building on its seven years' experience with the hardcore, General Electric is now pursuing a new alternative for some of its employees. The company's Chicago operations have negotiated a MA-4 contract with the Department of Labor to provide prevocational vestibule training with emphasis on motivation, attitudes, and follow-up support. Its goal is to help 400 hardcore disadvantaged people to become steady and satisfied workers.

In 1962, just prior to the large influx of black employees in the labor force, production employees, previously nonunion, chose the Sheet Metal Workers International Association, an old-line craft union, to represent them. SMWIA has a long racially discriminatory record in the building and metal trades. It first excluded Negroes from membership, then confined them to separate locals "with the consent of the white local" in the area involved, or else in "auxiliary locals" if consent of the white local is not obtained, with the Negro local under the supervision of the white local at all times.[76] Under pressure from state fair employment practice commissions and with the desire to expand into industrial plants, these rules were eliminated and Negroes admitted into locals, particularly industrial ones. In the building trades, however, locals of SMWIA have been repeatedly charged with discrimination and found to be discriminating by courts in several sectors of the country.[77] Even in its industrial locals the union attitude is best described as "laissez-faire."

That distrust should develop between the SMWIA local and its black constituents was nearly inevitable. In the summer of 1968 a Negro, Willie Plunkett, ran for local president, lost, and organized a group called the Afro-American Employees of Hotpoint. In October a black and a Puerto Rican employee engaged in a fight while at work, with the former being discharged as the instigator, and the latter suspended. The Afro-American group, by then known as the Black Federation of Labor, launched sitdowns, picketing, and other harassment which actually shut Hotpoint for one day. Plunkett and others were suspended, and it was explained to them that since SMWIA was the certified bargaining agent, the company could not deal with another union. Finally, the company sought a directive from the National Labor Relations Board proscribing further interference from the Black Federation.

In October 1969, a coalition of unions including SMWIA struck General Electric plants. A substantial number of employees including many blacks crossed picket lines and went to work before the four-months strike ended. Now, however, the Sheet Metal Workers local there has a black president and many black stewards, so that union integration is proceeding on a local basis in a quite different manner than appears to be typical in this union's construction locals.[78]

---

[76] Northrup, *Organized Labor and the Negro*, pp. 4-5.

[77] Los Angeles, New York, and St. Louis to name but three. The New York City experience is detailed in George D. Zuckerman, "The Sheet Metal Workers' Case: A Case History of Discrimination in the Building Trades," *Labor Law Journal*, Vol. LX (July 1969), pp. 416-427.

[78] Based upon briefs in noted NLRB case and interviews with company personnel. See also Victor Riesel's "Inside Labor" column issued October 25, 1968.

AFFIRMATIVE ACTION AND TERRORISTIC REACTION

Westinghouse has taken the lead in a great variety of programs in the communities in which it operates, besides greatly expanding its minority work force in recent years. Its hardcore training program in the Pittsburgh headquarters area has been written about elsewhere.[79] But perhaps nothing better illustrates its resolve to push integration than its reaction to a bombing in Vicksburg, Mississippi.

Westinghouse employed 660 persons, including 115 Negroes, in its Vicksburg light fixture plant in 1965. In late spring of that year, the plant commenced employing Negro women for the first time and began a campaign to employ "qualified Negro women for jobs as clerks, stenographers and secretaries."[80] About 4:15 a.m. on Saturday morning, June 19, 1965, someone tossed three ignited fire bombs into the garage of the industrial relations manager, damaging his two cars and the garage. Because he managed to put out the fires with a garden hose, the damage was contained.

Westinghouse's reaction was a model. Its director of corporate security and four other officials went immediately to Vicksburg from their Pittsburgh headquarters. The security director, a retired rear admiral, stated that the company "would not move its plant as the result of one incident." Then he added:

> But we want to make it clear that we are going to comply fully with government laws requiring equal employment opportunity from firms dealing with defense contracts. We can't throw away 20 percent of our business—it amounts to about half a billion a year in defense contracts alone—because of one plant.[81]

The plant manager, in a letter to employees written the Monday after the incident, stated:

> Westinghouse management immediately informed city, county, state, and Federal law enforcement officials, who moved swiftly into action in an effort to locate the person or persons behind this act. Who did this we don't know, but we expect to find out. We would have taken

[79] Theodore V. Purcell and Rosalind Webster, "Window on the Hard-Core World," *Harvard Business Review*, Vol. XLVII (July-August 1969), pp. 118-129; and *ibid.*, "196 Men Find a Chance," in Peter B. Doeringer, *Programs to Employ the Disadvantaged* (Englewood Cliffs, N.J.: Prentice-Hall, 1969), pp. 60-96.

[80] *New York Times*, June 24, 1965.

[81] *Ibid.*

similar vigorous action to protect the person and property of any Westinghouse employe attacked in this manner.

I want to make it crystal clear to everyone at this plant that Westinghouse and local management will not tolerate any actions that jeopardize the personal safety of employes.

All the law enforcement officials are cooperating to the fullest degree in efforts to track down who threw those fire bombs.

Justice and the rights of others must be respected, and we mean to do everything we can to see that they are.[82]

There was no repetition of the incident. Moreover, the vigorous reaction of Westinghouse and its clear threat to move a $30 million payroll out of Vicksburg forced the leadership of the community to choose between modern industry with integration, or lawlessness and no industry. The bombing was the fourth within a year. Three earlier ones damaged a civil rights meeting hall and two racially integrated restaurants. But it took the strong Westinghouse reaction to induce community and state action. The governor announced that "Mississippi will not tolerate incidents of this kind;"[83] the local papers editorialized about the impact of the possible loss of this and other plants that might locate there;[84] and the Chamber of Commerce issued the statement which is set forth in Figure 2 and had it printed in the local paper. As a local Negro newspaper reported, "The great wall of silence was finally broken"[85] when Westinghouse refused to be intimidated or to tolerate lawlessness.

UPGRADING THE LABOR FORCE

A major problem of integration is the upgrading of the labor force. Hiring persons for semiskilled jobs not only fails to meet the requirements of governmental compliance programs; something more is required to build a work force that will provide the skills necessary for effective operation. Two examples of Big Four activities in this regard are provided below.

Developing a Skilled Labor Force

American industry has for many years depended on foreign trained employees and their offspring to fill the key craftsman

---

[82] R. P. McFarlin, June 21, 1965. Copy in Industrial Research Unit files.

[83] *New York Times*, June 24, 1965.

[84] *Vicksburg Evening Post*, June 21 and 22, 1965.

[85] *Vicksburg Citizens' Appeal*, July 5, 1965.

FIGURE 2. *Statement of the Vicksburg Chamber of Commerce*

We all have been shocked and appalled by the cowardly firebombing of the home of Herman Vavra, and its serious implications.

We now call upon the entire community to join in a massive united effort to eliminate the kind of attitudes that could produce the Vavra incident.

The Vicksburg Chamber of Commerce has been open and forthright in its position advocating faithful support and obedience to law, including all phases of the Civil Rights Act. Those views are contained in a Statement of Policy adopted on February 9, 1965.

Thousands of responsible citizens, white and Negro, have shown their interest in having Warren County move ahead by encouraging industrial development and the progress it brings. We have been justly proud of results. The substantial upsurge of industrial progress is the fulfillment of a long-awaited promise. It has been deeply reflected in the spirit and enthusiasm of our citizenry.

The firebombing not only represented an unwarranted attack on innocent persons but also imperils that progress unless we wipe out its effects immediately by proving the community's desire to meet changing situations in a peaceful manner.

We need no further proof of the serious impact of this kind of violence than the fact Westinghouse would send a plane load of top officials to Vicksburg on a week-end to investigate circumstances that would produce an act of terror directed against the company's local personnel director.

An aroused public must aid in seeing that these unlawful acts cease and that persons responsible are brought to justice.

On July 2, Title VII of the 1964 Civil Rights Act goes into effect. This law requires many industries and other places of employment to wipe away every trace of racial discrimination in the conduct of their business.

The responsible citizens of Vicksburg and Warren County must see to it that these changes are carried out in a peaceful and orderly manner.

Employers coming under the Title VII provisions will have no choice. They must not discriminate against any minorities in hiring and firing, in advancement, or in plant activities.

This means that soon members of minority groups will be occupying positions in industry and business they have not occupied previously.

The public in Vicksburg and Warren County must provide the support to enable these industries and businesses to operate peacefully under the law soon to take effect.

THE CHAMBER OF COMMERCE URGES EACH RESPONSIBLE CITIZEN TO:

(1) If you are an employee in a plant or business coming under Title VII, cooperate in every way possible in whatever your employer MUST do to comply with this section of the Civil Rights Act.

(2) If you are a customer of a business that is complying with the Act, remember the businessman MUST follow the new Federal law if he is to remain in operation.

(3) If you have friends or relatives personally affected by the new law, give them your support and encouragement to see that Warren County makes a peaceful and orderly transition into this period of change.

Our community will not tolerate a repeat of any such heinous, criminal and unlawful act.

VICKSBURG CHAMBER OF COMMERCE

Adopted June 22, 1965

Source: *Vicksburg Evening Post,* June 23, 1965.

jobs in their plants. Today, however, the sons of immigrants are more likely to go to college than to take an apprenticeship. Moreover, plants which once located in areas in which this skilled labor supply also dwelled, now find themselves in all-black neighborhoods, where most residents have had little industrial experience or family backgrounds of craft work. Concerned with this problem, Western Electric has taken the lead in Newark, New Jersey in providing skills development programs. Working through the Newark Business and Industrial Coordinating Council, Western Electric had a primary role in organizing an on-the-job training program designed to qualify disadvantaged youth for machine shop work. First, classwork instruction provides shop math, blueprint reading, use of mechanical tools and measuring devices, etc. Then an eight weeks on-the-job program at Western Electric's Kearny works emphasizes drill press operation and includes layout work and drill sharpening as well.

Additionally, Western Electric has attempted to upgrade skills of the disadvantaged by establishing a small branch operation in Newark which makes products, but is primarily concerned with teaching work practices and machine shop operations to disadvantaged trainees by giving them actual industrial experience.[86] As employees become proficient they are transferred to the Kearny works. Some 300 have been transferred so far, with a 75 percent retention rate. A bonus of this operation has been supervisory training. Western Electric has found that by rotating supervisors through the Newark operation, it gives them a practical human relations course. The company believes this aids their management of the labor force which, by reason of company location, it must depend even more heavily in the future.[87]

The Newark program is but one of many Western Electric programs designed to increase opportunities for Negroes and to prepare them for work in its plants. Its conscious program of recruiting and upgrading Negroes is one of the most widespread and earliest conceived in the industry, and together with its

---

[86] For a description and evaluation of the Western Electric program in the Newark area, see Jack Chermik and Georgine Smith, "Employing the Disadvantaged," in Doeringer, *op. cit.*, pp. 14-23. For an earlier summary of the Western Electric equal opportunity programs, see R. G. Lawrence, "Western Electric's Commitment to Fair Employment," in Northrup and Rowan (eds.), *The Negro and Employment Opportunity*, pp. 137-145.

[87] Based upon interviews and field visits.

Chicago and Newark area locations, has helped to give this company one of the largest percentages of black employees in the industry. Upgrading has also proceeded rapidly so that today an increasing proportion of Negroes are found in white collar and lower management jobs. In the 1970's, Negroes will undoubtedly emerge in the upper ranks of the company.

Developing Service Technicians

Few will disagree that finding competent personnel to service appliances and radio and television sets—to the satisfaction of customers—has become a major problem for electrical manufacturing companies. Until recently, such employment was an almost exclusively white domain. Civil rights and labor market pressures have resulted in Negroes entering the trade, but whites still predominate.

RCA has pioneered in training disadvantaged personnel as television and electronic repair technicians in a number of cities throughout the country. Financed by U.S. Department of Labor training contracts, such training courses teach basic English and mathematics, as well as basic electronics and laboratory procedure. Since repairmen need driver's licenses, driver education is also provided as required. Moreover, in order to gain an idea of how the company operates, trainees work part-time in entry level positions associated with the operations for which they are being trained.

RCA has successfully trained over 500 persons for such jobs as TV antenna installation, customer repair of TV and "home entertainment units," and TV set dismantling and reassembly after being reworked in the shop. The course lasts for 18 months. RCA reports no serious difficulties in the reaction of customers to black servicemen.[88]

AFFIRMATIVE ACTION—FINAL COMMENT

The case stories noted here are not meant to be definitive or all-inclusive, but they do illustrate what the major firms are attempting to accomplish in the equal employment opportunity area. Obviously, much remains to be done. Achieving equality in electrical manufacturing is a long-term project that requires constant effort aimed at overcoming the effects of past neglect

---

88 Based upon interview.

in education and employment. Certainly a start has been made and progress is visible. The decade of the 1970's will require a continued effort if success is to be achieved.

UNION IMPACT

As we have already noted, the impact of unions on company racial policies has not been substantial. Companies do the hiring; union contracts help determine who is promoted or laid off, but we have found no evidence that unions have attempted either to discriminate against minorities, or to give them special advantages.

Another significant factor in the electrical manufacturing industry is that the unions involved, for the most part, are made up of strong locals which do not accept national union leadership automatically. Thus although the International Union of Electrical, Radio and Machine Workers (IUE) has taken a strong position against discrimination and has even acted as a contractor with the federal government to train disadvantaged employees for work in the industry, the policies of its local unions are more likely to reflect the realities of local union politics. For example, if there is a strong black minority among the membership, the local officials are more likely to be concerned with racial problems than if the minority membership is small.

Local union officials are subject to defeat if they stray too far from the wishes of their constituents. They can lead, but cannot get too far in front or back. Thus, as we shall discuss in the next section, the left-wing United Electrical, Radio and Machine Workers (UE) never negotiated a nondiscrimination contract with the Allen-Bradley Company of Milwaukee despite its long espousal of civil rights until the membership feared that contracts would be cancelled if company policies were not altered. Since the UE international did support equal opportunity, however, it was able to help work out a settlement of the controversy.

Where a company alters its policies from exclusion to integration, local union officials have difficulty accommodating to the new policies. Thus in a now integrated but formerly lily-white southern plant, the new Negro employees do not believe that the union committee pushes their grievances as hard as it does those of the white employees. The union local certainly did not oppose the former hiring policy and made no effort to smooth integration.[89] It is also likely that in such a situation, the new black employees would be suspicious of a white-dominated local until a real effort toward integrating them into the membership has been made. Although both international and especially local union policies can help or hinder integration at the plant level, our early conclusion that the unions in the electrical manufacturing industry are not a major factor in racial policies seems valid.

GOVERNMENTAL IMPACT

The fact that a significant share of electrical manufacturing and electronic research is performed for the United States government insures that governmental policies will have a profound and direct impact on the racial policies of the industry. The impact of past governmental policies has already been reviewed. Today, the electrical manufacturing industry is covered by the nondiscriminatory requirements of Title VII of the Civil Rights Act and similar laws in most states outside of the South. Judging by the general absence of court cases involving electrical manufacturing companies, one could suspect that complaints by individuals and activities by the Equal Employment Opportunity Commission and state agencies pursuant to Title VII and state laws have not been numerous. In fact, the companies in the industry tend to settle cases before or in the conciliation stage, thus avoiding litigation. Their relatively high demand for labor, carefully designed policies, and lack of overt discrimination all tend to reduce complaint and litigation procedures.

On the other hand, this does not mean that the Equal Employment Opportunity Commission is satisfied with the progress of the industry. Wherever, as in the case of electrical manufacturing, the percentage of minority employees is below that of industry generally or manufacturing industry, the EEOC is certain to be concerned. The "gap in understanding" that Professor Bernard E. Anderson noted between the management in public utilities and the EEOC is equally applicable in the electrical machinery industry:

> The commission, being an enforcement agency, looks to the statistical record for evidence that equal employment, in fact, exists in business firms. Corporate managers, however, often emphasize company progress in changing the focus, direction, and energy devoted to equal employment issues. Such changes within the business firm may not

[89] Interviews, Spring 1970.

result in significant numerical gains in the number of black workers employed or promoted during a one- or two-year period. The modification of traditional employment practices is no less real to corporate managers even though the results may not meet EEOC measurement criteria. To management, changes in the employment process that might result in long term gains for black workers are often considered more important than short term gains in numbers employed. The difference of opinion between EEOC and industry representatives regarding the meaning and relative importance of affirmative action programs and affirmative action results is one of the most difficult issues in the equal employment area today.[90]

More significant for the industry are the rules and regulations adopted pursuant to Executive Order No. 11246, particularly General Order No. 4 of the Office of Federal Contract Compliance. The Executive Order sets standards above and beyond Title VII which must be adhered to by government contractors.[91] General Order No. 4, issued on November 20, 1969, sets forth in great detail what affirmative action requires of companies. This includes: 1) an analysis of all major job classifications and an explanation why minorities may be underutilized; 2) the establishment of goals, targets and affirmative action commitments designed to relieve any shortcoming identified; and 3) the development and supply of data to government organizations, including not only racial-occupational figures to great detail, but also progression charts, seniority rosters, and an analysis of applicants by race for the various jobs.

Although General Order No. 4 requires "targets" or "goals" as measurements of "good faith efforts," the OFCC denies that "goals" and "quotas" are synonomous.[92] General Order No. 4 does, spell out for the first time what is meant by "affirmative action." The major electrical companies have all adopted policies to conform with General Order No. 4, and are in effect working to increase their proportion of black employees. This direct employment activity is in addition to various community activities,

---

90 Bernard E. Anderson, *Negro Employment in Public Utilities,* Studies of Negro Employment, Vol. III (Philadelphia: Industrial Research Unit, Wharton School of Finance and Commerce, University of Pennsylvania, 1970), p. 207.

91 The interesting constitutional question of whether the President can promulgate regulations pursuant to government procurement policies which go beyond, and may be inconsistent with, statutory law has never been tested.

92 Title VII of the Civil Rights Act of 1964 contains an express prohibition against the requirement of quotas.

special training endeavors or other social programs. Of course, the two are interacting. But goals are set and managers are being held accountable for results. The net effect will undoubtedly be a steady increase in the number and proportion of black employees, and a slower, but steady move toward a better balance of Negroes in the salaried and craftsmen positions.

The Allen-Bradley Case

A specific impact of Executive Order No. 11246 is illustrated by the situation at the Allen-Bradley Company in Milwaukee. In 1964, this company became engaged in a controversy over whether it was discriminatory. The issue was not overt discrimination, but the lack of minority employees which resulted from its policies of maintaining high employment standards and filling jobs by relying on referrals from existing employees. As a result, the labor force of the company remained 99.5 percent white.[93]

Pressed by the government, Allen-Bradley altered its standards, loosening requirements concerning Wisconsin residency and educational attainments, and began to advertise in minority papers for help. (It had not previously advertised for help anywhere.) Pushed further by government, which held a public hearing on whether it should be debarred from government contracts, Allen-Bradley began a steady buildup of its black employee work force from 35 in 1968 to 185 in 1969, and this has gone forward. With 6,500 employees, and a local population that is 10 percent black, this represented only a start toward fair employment, and did not satisfy the local minority organizations agitating and demonstrating for governmental sanctions against the company. Sanctions were rejected by the then Secretary of Labor, George P. Shultz (the case began under his predecessor W. Willard Wirtz), who instead accepted the company's progress, and proceeded by continuing governmental monitoring thereof.

An interesting sidelight in the case was the role of the Local No. 1111, one of the largest in the independent, left-wing United Electrical, Radio and Machine Workers. Idealogically committed to a pro-minority employee course, this local negotiated a nondiscrimination clause when the controversy over Allen-Bradley

[93] The full story of this case is found in the hearings and briefs of the proceeding initiated by the Office of Federal Contract Compliance. A good summary is in *Business Week*, October 12, 1968, pp. 122, 124, 126; and August 16, 1969, p. 44.

policies developed and then tried to steer a middle course between the wishes of its white members and its ideological position. It did exercise influence in both ways—bringing the membership to accept change and the government to accept the company's program.

The Allen-Bradley case is a good illustration of the impact of government procurement policy. It forced change that seemingly went beyond the requirements of Title VII of the Civil Rights Act to compel a reluctant employer to alter its policies, but stopped short of the drastic action of debarring a company. Debarment would force a court challenge of executive powers—something that all parties appear reluctant to do. More immediately, it would throw many persons out of work—a plant such as Allen-Bradley would probably close since debarment would not only prohibit the company from bidding on direct government contracts; it would also forbid other government contractors from doing business with a debarred company. The impact is so great that debarment has never been invoked. Yet it remains such a threat that no company cares to risk a conclusive showdown, especially over an issue that appears to place the company on the wrong side of social justice.

CONSUMER ORIENTATION AND COMPANY IMAGE

Another reason why companies in the electrical manufacturing industry are concerned with their racial policies and so cooperative with government is the strong consumer orientation of managements in such businesses as appliance manufacturing. These companies discovered the "Negro market" in the 1950's and began to develop ways to exploit it. Civil rights agitation in the 1960's caused them to reevaluate employment policies. Such companies are likely to try and keep abreast of public attitudes —neither too far in front nor behind—so as to appear forward looking but not extreme. Thus in the late 1950's and early 1960's, these companies worked to resolve southern school crises and to avoid locating plants where crises existed.

Western Electric, for example, does not sell to consumers, but as part of the Bell System, its executives have a deep concern for "company image," as a law abiding, socially responsible concern. Although it is difficult to know how much influence on racial policies consumer orientation and company image have, it is noteworthy that other studies in this series have found them to be significant,[94] and undoubtedly they contribute to management attitudes in electrical manufacturing. Certainly, they are a factor in the almost universal desire of the industry to avoid confrontation with the government on issues of minority employment and to present as attractive a posture as possible on this issue.

[94] See, e.g., Northrup, Rowan, *et al., Negro Employment in Basic Industry,* Part Eight, pp. 734-735.

CHAPTER VI

The Determinants of Industry Policy

Throughout this study we have discussed several factors affecting Negro employment and advancement in the electrical manufacturing industry. To summarize our study we shall now reemphasize these factors.

DEMAND FOR LABOR

There is no doubt that the need for labor has been an important factor in the employment of Negroes in the electrical industry. Negroes were first hired in substantial numbers during the acute labor shortage of World War II. Then the rapid growth of the industry in the fifties and sixties—employment more than doubled between 1950 and 1968—meant increased job opportunities for all, including Negroes. Where Negro labor was available, qualified, and aware of job opportunities in the electrical industry, Negroes were hired naturally with no conscious effort. In most cases, however, Negro employment did not expand until management deliberately tried to improve recruitment of black workers in the sixties. The need for labor did not in itself result in improved black hiring, but made it easier for management to carry out affirmative efforts with little threat to the white workers traditionally employed.

In some locations shortages in the labor market have recently dictated increased employment of Negroes, regardless of management policy. This is particularly true in a number of the large, older city plants where the racial composition of the neighborhood has changed and managers find that a large proportion, sometimes the vast majority, of their applicants are black. There is also some indication that some southern plants located in cities and towns with a large black population would have had difficulty in the late sixties finding well qualified applicants for semiskilled jobs had they not become less discriminatory in hiring.

The electrical industry continues to grow. The production of electrical energy has recently doubled every ten years and is expected to continue to do so. Electrical companies are planning for expansion. This growth in a labor intensive industry means that the electrical industry will continue to be in a good position to employ large numbers of Negroes. Increases will occur automatically in some areas and management is in an excellent position to implement affirmative action programs in many others.

Of course the electrical industry is not immune to recessions. There have been years of major declines in employment. Large companies were affected by the recession in the late fifties. Any cutbacks in employment will, of course, affect Negroes, and in some cases disproportionately because of their lack of seniority. Nevertheless, the history of the industry has been one of rapid recovery from setbacks and continued growth which is likely to offset any future temporary decline in employment.

JOB STRUCTURE

The electrical industry is technically oriented, with a high (37 percent) and growing proportion of white collar jobs. It also has a great number (38 percent) of semiskilled operative jobs for which relatively little training is necessary. The number of these jobs is also growing although the proportion is likely to be relatively static.

It is in unskilled and semiskilled jobs that hiring Negroes has been easiest, and it is here that the industry has made the most progress. In 1969 Negroes held 11.5 percent of the operative jobs and were overrepresented in the labor and service categories. The low point in blue collar employment was craftsmen where only 4.2 percent were black.

Much of the industry is featured by light manufacturing. Women are increasingly employed in electronic assembly work, and the greatest progress in blue collar jobs has been among Negro semiskilled employees. Male Negro progress has been much slower.

Integration of white collar employment is a greater challenge. Negro employment in clerical positions is low at 3.7 percent, but clerical training is relatively easy to supply. For the other white collar positions technical requirements limit the supply of qualified black applicants. Again, this has meant more progress for black females than for black males.

In short, the electrical job structure is relatively favorable to higher Negro employment, especially for women. But Negro craftsmen, technical people, and managers will come in much more slowly. The opportunities are there but the qualifications are demanding. To develop skilled cadres of blacks in the white collar jobs and in technical and management positions will require extensive investment of time and ingenuity on the part of society in general and industry in particular.

GOVERNMENT AND CIVIL RIGHTS PRESSURE

Government pressure has been a very important influence on racial employment policies in the electrical industry both directly and indirectly. In the 1950's all the major electrical firms had some plants investigated by state fair employment practices commissions. In the 1960's electrical companies, most of which are government contractors, began to be subject to extensive compliance reviews and findings at a number of plants were negative. Many companies responded by making more aggressive internal efforts in equal employment opportunity in the hope of avoiding closer federal and state supervision as well as in growing recognition of the urgency of the problem. Most important, as the federal government put increasing emphasis on the racial composition of the total work force as an important indicator of employment policy, many companies adopted a more systematic approach in internal evaluation. Individual company executives, to be sure, had held genuinely progressive positions before government efforts became strong, but full scale implementation came only after an increased government interest.

In the South, where equal employment ran counter to local mores, some plants had operated on a nondiscriminatory basis prior to President Kennedy's Executive Order of 1961 and the Civil Rights Act of 1964, but many others began to plan for full integration only in response to federal law. Plants in the South have moved ahead in total utilization of Negroes, but have lagged in employing those in white collar positions.

A closely related influence for change in the industry is the rapidly developing civil rights climate of the country. The activity of civil rights organizations and the riots in Newark, Detroit, Memphis, Chicago, Cleveland, and other cities led managers of electrical plants in these cities to develop programs for dealing with the total urban environment within which their plants are located. Few industries are so deeply involved in such programs as are companies in electrical manufacturing.

UNIONS AND SENIORITY

The major unions of the electrical industry have had little direct effect on Negro employment and upgrading. Employment is something handled by management rather than by the union, and upgrading is also largely in the hands of management. The International Union of Electrical, Radio and Machine Workers (IUE) and the United Electrical, Radio and Machine Workers (UE) have had a fairly active concern with civil rights and the advancement of Negro people, but their local unions have generally reacted in accordance with the wishes of the largely white membership. Other unions have been less interested or responsive. Most union-management contracts have long had a nondiscrimination clause, but these have neither been instigators of change, nor actively utilized except in a few instances.

Seniority requirements may be a moderate hindrance to Negro upgrading and promotion within the industry to the extent that they require the last hired to be upgraded last and laid off first in the event of cutbacks in employment. On the other hand, seniority is not rigid or narrow so that there are no racially motivated seniority bars, and seniority does give those with equal length of service the same opportunities. Hence seniority is neither a significant bar nor a spur to Negro advancement.

REGIONAL AND LOCATIONAL FACTORS

Location has had a significant impact on the racial composition of the electrical industry. The industry developed in manufacturing towns in the Northeast and Midwest at a time when there were almost no Negroes in the labor market there. The migration of Negroes to the North has greatly increased the number of Negroes employed in many large, older city plants, particularly in the Midwest and Middle Atlantic states. Many other northern plants, however, remain in virtually all white areas and traditional business considerations have led to the construction of new plants in suburban or small town areas where few Negroes dwell. Thus in the North an increasing number of Negroes will be employed in some areas, but location will continue to inhibit black employment somewhat unless management begins to apply affirmative action considerations to its choices of location. Given basic economics, this is not likely. Rather affirmative action is likely to be utilized to overcome locational disadvantages to minority employment.

Geographical diversification after World War II took the electrical industry into the South, but did not immediately bring improved opportunities for Negroes. In 1964 4.2 percent of southern electrical employees were black, a slightly higher percentage than that on the West Coast or in New England, but not very different from the Midwest and the Middle Atlantic states. With affirmative action and tight labor markets in the sixties, southern hiring practices changed and Negro employment spurted ahead of most other regions, although qualitatively it lagged. Moreover, the locational disadvantages of outlying locations and lack of public transportation are apparently being overcome to a large extent in the South.

MANAGERIAL POLICY

Managerial policy in the electrical industry in the past varied among different companies. A few notable leaders had a broad view—for example, David Sarnoff of RCA, A. B. Goetze of Western Electric, and Gerard Swope of General Electric—but these influences did not lead to aggressive activity toward hiring and upgrading minority people.

Prior to 1960, despite some executive eloquence and finely-cast policy statements, much of the electrical manufacturing industry had little to show in labor force integration. The record since 1960 has been better; indeed, more has been accomplished in the last ten years than was achieved in the previous fifty years. Since 1960, and especially since 1964, management policies have changed regarding the matter of corporate social responsibility, the reaction of the corporation to its environment, and efforts to deal with minority manpower. The major companies now all have extensive apparatus at their central offices to deal with equal employment opportunity issues, and local managers are under pressure and surveillance to advance minority employment. In the 1970's such programs should show increasing results if the industry can overcome the obstacles caused by the shortage of trained black technical employees.

Managerial policy in some cases has been affected by consumer orientation and concern with company image. To the extent that this continues, it is likely to be a positive factor.

CONCLUDING REMARKS

The electrical industry occupies a significant and important role in the American industrial scene. Both a part of and a contribution to our great technical apparatus, it has enjoyed growth and prosperity in most of the years of this century. Its concern with minority employment problems was not translated into effective programming until recent years because its demand for labor did not require the large numbers of unskilled labor that such industries as steel and automobiles needed so early in their development.

Now, spurred by government, the electrical manufacturing industry has developed policies, programs, and staff to expand opportunities for minorities at a rate faster than that of the past. *Business and Society,* a journal that is often critical of management efforts pertaining to "social responsibility," has termed General Electric's equal employment program "the most comprehensive one of its kind that we know of in the corporate world." [95] The strong efforts of the other major companies have likewise drawn much deserved praise. Their success and that of the other companies in the industry will determine whether equal employment can be achieved in an industry with high education and technical requirements. The electrical industry may now have a greater opportunity than almost any other industry in the country to make notable and creative experimental advances in finding ways to integrate black people with the labor force. Whether it will happen may well depend on the extent and durability of its management commitment and initiative.

[95] *Business and Society,* Vol. 3, No. 12 (December 17, 1970), p. 1.

Appendix

Basic Statistical Tables, 1964, 1966, 1967, and 1969

Table A-1. *Electrical Manufacturing Industry Employment by Race, Sex, and Occupational Group 1,588 Establishments, United States, 1964*

Occupational Group	All Employees			Male			Female		
	Total	Negro	Percent Negro	Total	Negro	Percent Negro	Total	Negro	Percent Negro
Officials and managers	74,988	193	0.3	73,510	186	0.3	1,478	7	0.5
Professionals	114,314	660	0.6	112,353	631	0.6	1,961	29	1.5
Technicians	76,346	1,300	1.7	70,983	1,182	1.7	5,363	118	2.2
Sales workers	14,586	56	0.4	12,983	24	0.2	1,603	32	2.0
Office and clerical	126,258	2,228	1.8	41,979	775	1.8	84,279	1,453	1.7
Total white collar	406,492	4,437	1.1	311,808	2,798	0.9	94,684	1,639	1.7
Craftsmen	123,598	2,587	2.1	116,019	2,039	1.8	7,579	548	7.2
Operatives	395,825	24,235	6.1	186,118	10,937	5.9	209,707	13,298	6.3
Laborers	86,303	5,230	6.1	46,910	3,331	7.1	39,393	1,899	4.8
Service workers	15,760	2,554	16.2	13,861	2,210	15.9	1,899	344	18.1
Total blue collar	621,486	34,606	5.6	362,908	18,517	5.1	258,578	16,089	6.2
Total	1,027,978	39,043	3.8	674,716	21,315	3.2	353,262	17,728	5.0

Source: Data in authors' possession.

TABLE A-2. *Electrical Manufacturing Industry Employment by Race, Sex, and Occupational Group 2,901 Establishments, United States, 1966*

Occupational Group	All Employees			Male			Female		
	Total	Negro	Percent Negro	Total	Negro	Percent Negro	Total	Negro	Percent Negro
Officials and managers	117,227	509	0.4	114,445	458	0.4	2,782	51	1.8
Professionals	175,987	1,282	0.7	172,367	1,208	0.7	3,620	74	2.0
Technicians	113,008	2,508	2.2	104,562	2,291	2.2	8,446	217	2.6
Sales workers	21,232	53	0.2	19,273	30	0.2	1,959	23	1.2
Office and clerical	203,179	3,969	2.0	66,699	1,559	2.3	136,480	2,410	1.8
Total white collar	630,633	8,321	1.3	477,346	5,546	1.2	153,287	2,775	1.8
Craftsmen	203,873	6,057	3.0	190,185	5,016	2.6	13,688	1,041	7.6
Operatives	689,789	57,672	8.4	311,308	26,361	8.5	378,481	31,311	8.3
Laborers	155,546	15,439	9.9	72,594	8,281	11.4	82,952	7,158	8.6
Service workers	26,389	4,626	17.5	22,648	4,037	17.8	3,741	589	15.7
Total blue collar	1,075,597	83,794	7.8	596,735	43,695	7.3	478,862	40,099	8.4
Total	1,706,230	92,115	5.4	1,074,081	49,241	4.6	632,149	42,874	6.8

Source: U.S. Equal Employment Opportunity Commission, *Job Patterns for Minorities and Women in Private Industry, 1966,* Report No. 1 (Washington: The Commission, 1968), Part II.

TABLE A-3. *Electrical Manufacturing Industry Employment by Race, Sex, and Occupational Group 3,436 Establishments, United States, 1967*

Occupational Group	All Employees			Male			Female		
	Total	Negro	Percent Negro	Total	Negro	Percent Negro	Total	Negro	Percent Negro
Officials and managers	134,885	842	0.6	131,720	764	0.6	3,165	78	2.5
Professionals	195,542	1,640	0.8	191,118	1,554	0.8	4,424	86	1.9
Technicians	132,836	3,568	2.7	121,955	3,159	2.6	10,881	409	3.8
Sales workers	23,682	55	0.2	21,722	40	0.2	1,960	15	0.8
Office and clerical	229,935	5,515	2.4	73,345	2,202	3.0	156,590	3,313	2.1
Total white collar	716,880	11,620	1.6	539,860	7,719	1.4	177,020	3,901	2.2
Craftsmen	221,942	8,422	3.8	205,117	6,734	3.3	16,825	1,688	10.0
Operatives	792,630	78,758	9.9	344,628	33,508	9.7	448,002	45,250	10.1
Laborers	172,411	20,499	11.9	76,128	10,608	13.9	96,283	9,891	10.3
Service workers	27,997	5,167	18.5	24,026	4,493	18.7	3,971	674	17.0
Total blue collar	1,214,980	122,846	9.3	649,899	55,343	8.5	565,081	57,503	10.2
Total	1,931,860	124,466	6.4	1,189,759	63,062	5.3	742,101	61,404	8.3

Source: U.S. Equal Employment Opportunity Commission, *Job Patterns for Minorities and Women in Private Industry, 1967*, Report No. 2 (Washington: The Commission, 1970), Vol. 1.

TABLE A-4. *Electrical Manufacturing Industry*
Employment by Race, Sex, and Occupational Group
3,533 Establishments, United States, 1969

Occupational Group	All Employees			Male			Female		
	Total	Negro	Percent Negro	Total	Negro	Percent Negro	Total	Negro	Percent Negro
Officials and managers	137,731	1,265	0.9	134,087	1,136	0.8	3,644	129	3.5
Professionals	176,851	1,844	1.0	171,489	1,745	1.0	5,362	99	1.8
Technicians	124,664	3,921	3.1	113,766	3,471	3.1	10,898	450	4.1
Sales workers	24,438	158	0.6	21,724	114	0.5	2,714	44	1.6
Office and clerical	211,348	8,243	3.9	63,058	2,750	4.4	148,290	5,493	3.7
Total white collar	675,032	15,431	2.3	504,124	9,216	1.8	170,908	6,215	3.6
Craftsmen	210,937	8,861	4.2	194,336	7,166	3.7	16,601	1,695	10.2
Operatives	664,749	76,593	11.5	289,418	31,742	11.0	375,331	44,851	11.9
Laborers	169,545	21,616	12.7	77,027	10,646	13.8	92,518	10,970	11.9
Service workers	27,411	5,253	19.2	23,316	4,525	19.4	4,095	728	17.8
Total blue collar	1,072,642	112,323	10.5	584,097	54,079	9.3	488,545	58,244	11.9
Total	1,747,674	127,754	7.3	1,088,221	63,295	5.8	659,453	64,459	9.8

Source: Preliminary Equal Employment Opportunity Commission data, 1969.

Note: This table is based on all employment reported to the EEOC. Regional tables are computed from data only for those states with 10 or more establishments or with 5 establishments and at least 2,000 employees.

TABLE A-5. *Electrical Manufacturing Industry Employment by Race, Sex, and Occupational Group 211 Establishments, New England Region, 1964*

Occupational Group	All Employees			Male			Female		
	Total	Negro	Percent Negro	Total	Negro	Percent Negro	Total	Negro	Percent Negro
Officials and managers	7,047	13	0.2	6,902	10	0.1	145	3	2.1
Professionals	13,726	88	0.6	13,404	84	0.6	322	4	1.2
Technicians	8,740	103	1.2	8,266	99	1.2	474	4	0.8
Sales workers	2,032	15	0.7	1,777	3	0.2	255	12	4.7
Office and clerical	13,842	107	0.8	4,143	40	1.0	9,699	67	0.7
Total white collar	45,387	326	0.7	34,492	236	0.7	10,895	90	0.8
Craftsmen	13,143	255	1.9	12,166	148	1.2	977	107	11.0
Operatives	43,202	1,480	3.4	16,252	455	2.8	26,950	1,025	3.8
Laborers	8,885	492	5.5	2,429	137	5.6	6,456	355	5.5
Service workers	1,754	122	7.0	1,564	108	6.9	190	14	7.4
Total blue collar	66,984	2,349	3.5	32,411	848	2.6	34,573	1,501	4.3
Total	112,371	2,675	2.4	66,903	1,084	1.6	45,468	1,591	3.5

Source: Data in authors' possession.

TABLE A-6. *Electrical Manufacturing Industry Employment by Race, Sex, and Occupational Group 298 Establishments, New England Region, 1966*

Occupational Group	All Employees			Male			Female		
	Total	Negro	Percent Negro	Total	Negro	Percent Negro	Total	Negro	Percent Negro
Officials and managers	9,776	23	0.2	9,551	20	0.2	225	3	1.3
Professionals	16,105	84	0.5	15,820	84	0.5	285	—	—
Technicians	10,345	166	1.6	9,654	157	1.6	691	9	1.3
Sales workers	2,430	5	0.2	2,139	3	0.1	291	2	0.7
Office and clerical	19,316	184	1.0	6,087	67	1.1	13,229	117	0.9
Total white collar	57,972	462	0.8	43,251	331	0.8	14,721	131	0.9
Craftsmen	17,033	417	2.4	15,864	263	1.7	1,169	154	13.2
Operatives	62,360	2,578	4.1	23,792	1,010	4.2	38,568	1,568	4.1
Laborers	13,172	1,090	8.3	3,551	276	7.8	9,621	814	8.5
Service workers	2,395	131	5.5	2,226	119	5.3	169	12	7.1
Total blue collar	94,960	4,216	4.4	45,433	1,668	3.7	49,527	2,548	5.1
Total	152,932	4,678	3.1	88,684	1,999	2.3	64,248	2,679	4.2

Source: U.S. Equal Employment Opportunity Commission, *Job Patterns for Minorities and Women in Private Industry, 1966*, Report No. 1 (Washington: The Commission, 1968), Part II.

TABLE A-7. *Electrical Manufacturing Industry Employment by Race, Sex, and Occupational Group 344 Establishments, New England Region, 1967*

Occupational Group	All Employees			Male			Female		
	Total	Negro	Percent Negro	Total	Negro	Percent Negro	Total	Negro	Percent Negro
Officials and managers	10,960	51	0.5	10,692	45	0.4	268	6	2.2
Professionals	16,358	95	0.6	15,980	92	0.6	378	3	0.8
Technicians	10,636	179	1.7	9,755	163	1.7	881	16	1.8
Sales workers	2,316	4	0.2	2,005	2	0.1	311	2	0.6
Office and clerical	19,888	251	1.3	5,930	82	1.4	13,958	169	1.2
Total white collar	60,158	580	1.0	44,362	384	0.9	15,796	196	1.2
Craftsmen	17,681	528	3.0	16,543	348	2.1	1,138	180	15.8
Operatives	69,066	4,039	5.8	24,555	1,327	5.4	44,511	2,712	6.1
Laborers	12,659	956	7.6	2,992	333	11.1	9,667	623	6.4
Service workers	2,204	158	7.2	1,991	137	6.9	213	21	9.9
Total blue collar	101,610	5,681	5.6	46,081	2,145	4.7	55,529	3,536	6.4
Total	161,768	6,261	3.9	90,443	2,529	2.8	71,325	3,732	5.2

Source: U.S. Equal Employment Opportunity Commission, *Job Patterns for Minorities and Women in Private Industry, 1967,* Report No. 2 (Washington: The Commission, 1970), Vol. 1.

TABLE A-8. *Electrical Manufacturing Industry*
Employment by Race, Sex, and Occupational Group
341 Establishments, New England Region, 1969

Occupational Group	All Employees			Male			Female		
	Total	Negro	Percent Negro	Total	Negro	Percent Negro	Total	Negro	Percent Negro
Officials and managers	10,615	65	0.6	10,298	56	0.5	317	9	2.8
Professionals	16,004	132	0.8	15,507	128	0.8	497	4	0.8
Technicians	9,575	256	2.7	8,720	220	2.5	855	36	4.2
Sales workers	2,311	11	0.5	1,881	8	0.4	430	3	0.7
Office and clerical	16,999	362	2.1	4,837	115	2.4	12,162	247	2.0
Total white collar	55,504	826	1.5	41,243	527	1.3	14,261	299	2.1
Craftsmen	13,809	492	3.6	12,612	356	2.8	1,197	136	11.4
Operatives	54,433	3,116	5.7	18,778	1,099	5.9	35,655	2,017	5.7
Laborers	12,563	807	6.4	3,005	285	9.5	9,558	522	5.5
Service workers	2,428	164	6.8	2,057	142	6.9	371	22	5.9
Total blue collar	83,233	4,579	5.5	36,452	1,882	5.2	46,781	2,697	5.8
Total	138,737	5,405	3.9	77,695	2,409	3.1	61,042	2,996	4.9

Source: Preliminary Equal Employment Opportunity Commission data, 1969.

TABLE A-9. *Electrical Manufacturing Industry Employment by Race, Sex, and Occupational Group 454 Establishments, Middle Atlantic Region, 1964*

Occupational Group	All Employees			Male			Female		
	Total	Negro	Percent Negro	Total	Negro	Percent Negro	Total	Negro	Percent Negro
Officials and managers	26,122	95	0.4	25,512	93	0.4	610	2	0.3
Professionals	38,227	289	0.8	37,546	275	0.7	681	14	2.1
Technicians	24,914	535	2.1	23,408	499	2.1	1,506	36	2.4
Sales workers	3,797	22	0.6	3,373	9	0.3	424	13	3.1
Office and clerical	43,903	956	2.2	15,513	353	2.3	28,390	603	2.1
Total white collar	136,963	1,897	1.4	105,352	1,229	1.2	31,611	668	2.1
Craftsmen	40,154	1,064	2.6	38,594	909	2.4	1,560	155	9.9
Operatives	100,881	7,958	7.9	50,740	3,066	6.0	50,141	4,892	9.8
Laborers	22,319	1,573	7.0	11,040	785	7.1	11,279	788	7.0
Service workers	4,966	569	11.5	4,238	483	11.4	728	86	11.8
Total blue collar	168,320	11,164	6.6	104,612	5,243	5.0	63,708	5,921	9.3
Total	305,283	13,061	4.3	209,964	6,472	3.1	95,319	6,589	6.9

Source: Data in authors' possession.

TABLE A-10. *Electrical Manufacturing Industry Employment by Race, Sex, and Occupational Group 689 Establishments, Middle Atlantic Region, 1966*

Occupational Group	All Employees			Male			Female		
	Total	Negro	Percent Negro	Total	Negro	Percent Negro	Total	Negro	Percent Negro
Officials and managers	32,848	213	0.6	32,007	184	0.6	841	29	3.4
Professionals	43,507	364	0.8	42,711	355	0.8	796	9	1.1
Technicians	31,493	888	2.8	29,306	814	2.8	2,187	74	3.4
Sales workers	5,248	21	0.4	4,737	11	0.2	511	10	2.0
Office and clerical	53,837	1,477	2.7	18,704	548	2.9	35,133	929	2.6
Total white collar	166,933	2,963	1.8	127,465	1,912	1.5	39,468	1,051	2.7
Craftsmen	55,692	1,864	3.3	53,254	1,701	3.2	2,438	163	6.7
Operatives	147,971	16,079	10.9	67,990	6,228	9.2	79,981	9,851	12.3
Laborers	27,711	3,695	13.3	13,249	1,820	13.7	14,462	1,875	13.0
Service workers	7,014	996	14.2	5,794	864	14.9	1,220	132	10.8
Total blue collar	238,388	22,634	9.5	140,287	10,613	7.6	98,101	12,021	12.3
Total	405,321	25,597	6.3	267,752	12,525	4.7	137,569	13,072	9.5

Source: U.S. Equal Employment Opportunity Commission, *Job Patterns for Minorities and Women in Private Industry, 1966*, Report No. 1 (Washington: The Commission, 1968), Part II.

TABLE A-11. *Electrical Manufacturing Industry Employment by Race, Sex, and Occupational Group 829 Establishments, Middle Atlantic Region, 1967*

Occupational Group	All Employees			Male			Female		
	Total	Negro	Percent Negro	Total	Negro	Percent Negro	Total	Negro	Percent Negro
Officials and managers	34,427	271	0.8	33,689	233	0.7	738	38	5.1
Professionals	44,039	434	1.0	43,052	410	1.0	987	24	2.4
Technicians	33,293	1,219	3.7	30,690	1,025	3.3	2,603	194	7.5
Sales workers	5,712	27	0.5	5,215	18	0.3	497	9	1.8
Office and clerical	57,899	1,837	3.2	19,704	701	3.6	38,195	1,136	3.0
Total white collar	175,370	3,788	2.2	132,350	2,387	1.8	43,020	1,401	3.3
Craftsmen	60,700	2,661	4.4	57,830	2,221	3.8	2,870	440	15.3
Operatives	170,628	20,234	11.9	80,115	7,917	9.9	90,513	12,317	13.6
Laborers	38,057	4,723	12.4	17,921	2,282	12.7	20,136	2,441	12.1
Service workers	7,276	1,052	14.5	6,181	921	14.9	1,095	131	12.0
Total blue collar	276,661	28,670	10.4	162,047	13,341	8.2	114,614	15,329	13.4
Total	452,031	32,458	7.2	294,397	15,728	5.3	157,634	16,730	10.6

Source: U.S. Equal Employment Opportunity Commission, *Job Patterns for Minorities and Women in Private Industry, 1967*, Report No. 2 (Washington: The Commission, 1970), Vol. 1.

TABLE A-12. *Electrical Manufacturing Industry Employment by Race, Sex, and Occupational Group 847 Establishments, Middle Atlantic Region, 1969*

Occupational Group	All Employees			Male			Female		
	Total	Negro	Percent Negro	Total	Negro	Percent Negro	Total	Negro	Percent Negro
Officials and managers	37,056	408	1.1	36,241	361	1.0	815	47	5.8
Professionals	45,512	555	1.2	44,222	526	1.2	1,290	29	2.2
Technicians	32,385	1,323	4.1	29,888	1,199	4.0	2,497	124	5.0
Sales workers	6,164	38	0.6	5,463	23	0.4	701	15	2.1
Office and clerical	56,468	2,688	4.8	18,083	873	4.8	38,385	1,815	4.7
Total white collar	177,585	5,012	2.8	133,897	2,982	2.2	43,688	2,030	4.6
Craftsmen	62,789	3,118	5.0	59,724	2,610	4.4	3,065	508	16.6
Operatives	142,793	17,290	12.1	70,449	6,892	9.8	72,344	10,398	14.4
Laborers	36,053	4,432	12.3	17,531	2,054	11.7	18,522	2,378	12.8
Service workers	7,662	1,254	16.4	6,695	1,104	16.5	967	150	15.5
Total blue collar	249,297	26,094	10.5	154,399	12,660	8.2	94,898	13,434	14.2
Total	426,882	31,106	7.3	288,296	15,642	5.4	138,586	15,464	11.2

Source: Preliminary Equal Employment Opportunity Commission data, 1969.

TABLE A-13. *Electrical Manufacturing Industry Employment by Race, Sex, and Occupational Group 409 Establishments, Midwest Region, 1964*

Occupational Group	All Employees			Male			Female		
	Total	Negro	Percent Negro	Total	Negro	Percent Negro	Total	Negro	Percent Negro
Officials and managers	23,318	42	0.2	22,906	41	0.2	412	1	0.2
Professionals	25,287	81	0.3	24,918	79	0.3	369	2	0.5
Technicians	17,141	177	1.0	15,882	162	1.0	1,259	15	1.2
Sales workers	6,132	6	0.1	5,487	4	0.1	645	2	0.3
Office and clerical	36,119	604	1.7	12,394	193	1.6	23,725	411	1.7
Total white collar	107,997	910	0.8	81,587	479	0.6	26,410	431	1.6
Craftsmen	44,700	867	1.9	41,775	629	1.5	2,925	238	8.1
Operatives	164,774	9,926	6.0	86,786	5,357	6.2	77,988	4,569	5.9
Laborers	34,536	1,795	5.2	21,886	1,373	6.3	12,650	422	3.3
Service workers	5,869	947	16.1	5,106	782	15.3	763	165	21.6
Total blue collar	249,879	13,535	5.4	155,553	8,141	5.2	94,326	5,394	5.7
Total	357,876	14,445	4.0	237,140	8,620	3.6	120,736	5,825	4.8

Source: Data in authors' possession.

TABLE A-14. *Electrical Manufacturing Industry*
Employment by Race, Sex, and Occupational Group
854 Establishments, Midwest Region, 1966

Occupational Group	All Employees			Male			Female		
	Total	Negro	Percent Negro	Total	Negro	Percent Negro	Total	Negro	Percent Negro
Officials and managers	36,945	136	0.4	36,043	124	0.3	902	12	1.3
Professionals	33,849	168	0.5	33,343	142	0.4	506	26	5.1
Technicians	26,460	503	1.9	24,576	458	1.9	1,884	45	2.4
Sales workers	8,255	16	0.2	7,557	9	0.1	698	7	1.0
Office and clerical	61,154	1,050	1.7	21,177	350	1.7	39,977	700	1.8
Total white collar	166,663	1,873	1.1	122,696	1,083	0.9	43,967	790	1.8
Craftsmen	68,220	2,056	3.0	64,425	1,603	2.5	3,795	453	11.9
Operatives	293,102	25,490	8.7	142,621	12,844	9.0	150,481	12,646	8.4
Laborers	75,910	7,091	9.3	34,458	3,736	10.8	41,452	3,355	8.1
Service workers	9,607	1,558	16.2	8,062	1,320	16.4	1,545	238	15.4
Total blue collar	446,839	36,195	8.1	249,566	19,503	7.8	197,273	16,692	8.5
Total	613,502	38,068	6.2	372,262	20,586	5.5	241,240	17,482	7.2

Source: U.S. Equal Employment Opportunity Commission, *Job Patterns for Minorities and Women in Private Industry, 1966*, Report No. 1 (Washington: The Commission, 1968), Part II.

TABLE A-15. *Electrical Manufacturing Industry Employment by Race, Sex, and Occupational Group 999 Establishments, Midwest Region, 1967*

Occupational Group	All Employees			Male			Female		
	Total	Negro	Percent Negro	Total	Negro	Percent Negro	Total	Negro	Percent Negro
Officials and managers	41,656	304	0.7	40,551	281	0.7	1,105	23	2.1
Professionals	37,587	291	0.8	36,863	273	0.7	724	18	2.5
Technicians	30,252	762	2.5	28,022	683	2.4	2,230	79	3.5
Sales workers	9,366	13	0.1	8,637	9	0.1	729	4	0.5
Office and clerical	67,550	1,462	2.2	22,247	468	2.1	45,303	994	2.2
Total white collar	186,411	2,832	1.5	136,320	1,714	1.3	50,091	1,118	2.2
Craftsmen	73,130	2,965	4.1	67,153	2,162	3.2	5,977	803	13.4
Operatives	321,976	30,653	9.5	148,773	14,178	9.5	173,203	16,475	9.5
Laborers	78,341	9,067	11.6	33,436	4,883	14.6	44,905	4,184	9.3
Service workers	10,089	1,624	16.1	8,244	1,331	16.1	1,845	293	15.9
Total blue collar	483,536	44,309	9.2	257,606	22,554	8.8	225,930	21,755	9.6
Total	669,947	47,141	7.0	393,926	24,268	6.2	276,021	22,873	8.3

Source: U.S. Equal Employment Opportunity Commission, *Job Patterns for Minorities and Women in Private Industry, 1967*, Report No. 2 (Washington: The Commission, 1970), Vol. I.

TABLE A-16. *Electrical Manufacturing Industry*
Employment by Race, Sex, and Occupational Group
1,003 Establishments, Midwest Region, 1969

Occupational Group	All Employees			Male			Female		
	Total	Negro	Percent Negro	Total	Negro	Percent Negro	Total	Negro	Percent Negro
Officials and managers	40,582	439	1.1	39,444	392	1.0	1,138	47	4.1
Professionals	34,755	366	1.1	33,674	332	1.0	1,081	34	3.1
Technicians	28,987	804	2.8	26,849	714	2.7	2,138	90	4.2
Sales workers	8,716	81	0.9	8,058	65	0.8	658	16	2.4
Office and clerical	61,367	2,679	4.4	17,820	706	4.0	43,547	1,973	4.5
Total white collar	174,407	4,369	2.5	125,845	2,209	1.8	48,562	2,160	4.4
Craftsmen	66,034	2,875	4.4	61,205	2,197	3.6	4,829	678	14.0
Operatives	248,379	25,678	10.3	114,109	11,971	10.5	134,270	13,707	10.2
Laborers	73,949	8,494	11.5	34,656	4,713	13.6	39,293	3,781	9.6
Service workers	9,197	1,442	15.7	7,317	1,171	16.0	1,880	271	14.4
Total blue collar	397,559	38,489	9.7	217,287	20,052	9.2	180,272	18,437	10.2
Total	571,966	42,858	7.5	343,132	22,261	6.5	228,834	20,597	9.0

Source: Preliminary Equal Employment Opportunity Commission data, 1969.

TABLE A-17. *Electrical Manufacturing Industry Employment by Race, Sex, and Occupational Group 253 Establishments, West Coast Region, 1964*

Occupational Group	All Employees			Male			Female		
	Total	Negro	Percent Negro	Total	Negro	Percent Negro	Total	Negro	Percent Negro
Officials and managers	8,758	31	0.4	8,607	31	0.4	151	—	—
Professionals	17,162	140	0.8	16,761	136	0.8	401	4	1.0
Technicians	13,822	352	2.5	12,779	309	2.4	1,043	43	4.1
Sales workers	1,409	4	0.3	1,233	—	—	176	4	2.3
Office and clerical	14,647	184	1.3	3,835	88	2.3	10,812	96	0.9
Total white collar	55,798	711	1.3	43,215	564	1.3	12,583	147	1.2
Craftsmen	10,956	221	2.0	9,574	180	1.9	1,382	41	3.0
Operatives	25,191	1,206	4.8	9,069	358	3.9	16,122	848	5.3
Laborers	3,293	121	3.7	1,771	100	5.6	1,522	21	1.4
Service workers	1,145	258	22.5	1,040	243	23.4	105	15	14.3
Total blue collar	40,585	1,806	4.4	21,454	881	4.1	19,131	925	4.8
Total	96,383	2,517	2.6	64,669	1,445	2.2	31,714	1,072	3.4

Source: Data in authors' possession.

TABLE A-18. *Electrical Manufacturing Industry Employment by Race, Sex, and Occupational Group 477 Establishments, West Coast Region, 1966*

Occupational Group	All Employees			Male			Female		
	Total	Negro	Percent Negro	Total	Negro	Percent Negro	Total	Negro	Percent Negro
Officials and managers	21,181	106	0.5	20,798	102	0.5	383	4	1.0
Professionals	54,365	505	0.9	52,806	490	0.9	1,559	15	1.0
Technicians	27,841	666	2.4	25,422	624	2.5	2,419	42	1.7
Sales workers	3,124	8	0.3	2,814	6	0.2	310	2	0.6
Office and clerical	39,868	734	1.8	11,235	375	3.3	28,633	359	1.3
Total white collar	146,379	2,019	1.4	113,075	1,597	1.4	33,304	422	1.3
Craftsmen	33,591	1,133	3.4	29,938	936	3.1	3,653	197	5.4
Operatives	64,226	5,106	8.0	27,623	2,132	7.7	36,603	2,974	8.1
Laborers	9,093	865	9.5	4,793	586	12.2	4,300	279	6.5
Service workers	3,252	786	24.2	2,973	704	23.7	279	82	29.4
Total blue collar	110,162	7,890	7.2	65,327	4,358	6.7	44,835	3,532	7.9
Total	256,541	9,909	3.9	178,402	5,955	3.3	78,139	3,954	5.1

Source: U.S. Equal Employment Opportunity Commission, *Job Patterns for Minorities and Women in Private Industry, 1966*, Report No. 1 (Washington: The Commission, 1968), Part II.

TABLE A-19. *Electrical Manufacturing Industry Employment by Race, Sex, and Occupational Group 522 Establishments, West Coast Region, 1967*

Occupational Group	All Employees			Male			Female		
	Total	Negro	Percent Negro	Total	Negro	Percent Negro	Total	Negro	Percent Negro
Officials and managers	26,618	147	0.6	26,056	140	0.5	562	7	1.2
Professionals	62,088	615	1.0	60,498	594	1.0	1,590	21	1.3
Technicians	31,575	830	2.6	28,617	783	2.7	2,958	47	1.6
Sales workers	3,380	2	0.1	3,124	2	0.1	256	—	—
Office and clerical	48,369	1,086	2.2	13,648	541	4.0	34,721	545	1.6
Total white collar	172,030	2,680	1.6	131,943	2,060	1.6	40,087	620	1.5
Craftsmen	36,655	1,146	3.1	32,419	979	3.0	4,236	167	3.9
Operatives	76,042	6,544	8.6	32,958	3,265	9.9	43,084	3,279	7.6
Laborers	10,844	1,141	10.5	4,888	532	10.9	5,956	609	10.2
Service workers	3,950	957	24.2	3,646	880	24.1	304	77	25.3
Total blue collar	127,491	9,788	7.7	73,911	5,656	7.7	53,580	4,132	7.7
Total	299,521	12,468	4.2	205,854	7,716	3.7	93,667	4,752	5.1

Source: U.S. Equal Employment Opportunity Commission, *Job Patterns for Minorities and Women in Private Industry, 1967,* Report No. 2 (Washington: The Commission, 1970), Vol. I.

TABLE A-20. *Electrical Manufacturing Industry Employment by Race, Sex, and Occupational Group 548 Establishments, West Coast Region, 1969*

Occupational Group	All Employees			Male			Female		
	Total	Negro	Percent Negro	Total	Negro	Percent Negro	Total	Negro	Percent Negro
Officials and managers	22,486	193	0.9	21,954	185	0.8	532	8	1.5
Professionals	42,391	484	1.1	40,898	467	1.1	1,493	17	1.1
Technicians	25,364	723	2.9	22,765	671	2.9	2,599	52	2.0
Sales workers	3,363	13	0.4	3,021	9	0.3	342	4	1.2
Office and clerical	34,277	888	2.6	8,594	404	4.7	25,683	484	1.9
Total white collar	127,881	2,301	1.8	97,232	1,736	1.8	30,649	565	1.8
Craftsmen	27,681	826	3.0	23,250	605	2.6	4,431	221	5.0
Operatives	56,326	4,411	7.8	23,628	2,037	8.6	32,698	2,374	7.3
Laborers	12,095	1,121	9.3	4,752	500	10.5	7,343	621	8.5
Service workers	2,626	567	21.6	2,374	524	22.1	252	43	17.1
Total blue collar	98,728	6,925	7.0	54,004	3,666	6.8	44,724	3,259	7.3
Total	226,609	9,226	4.1	151,236	5,402	3.6	75,373	3,824	5.1

Source: Preliminary Equal Employment Opportunity Commission data, 1969.

TABLE A-21. *Electrical Manufacturing Industry Employment by Race, Sex, and Occupational Group 235 Establishments, South Region, 1964*

Occupational Group	All Employees			Male			Female		
	Total	Negro	Percent Negro	Total	Negro	Percent Negro	Total	Negro	Percent Negro
Officials and managers	9,392	10	0.1	9,240	10	0.1	152	—	—
Professionals	19,397	61	0.3	19,223	56	0.3	174	5	2.9
Technicians	11,087	129	1.2	10,047	109	1.1	1,040	20	1.9
Sales workers	1,175	4	0.3	1,072	3	0.3	103	1	1.0
Office and clerical	17,081	370	2.2	5,906	97	1.6	11,175	273	2.4
Total white collar	58,132	574	1.0	45,488	275	0.6	12,644	299	2.4
Craftsmen	14,179	177	1.2	13,502	170	1.3	677	7	1.0
Operatives	59,405	3,634	6.1	22,691	1,693	7.5	36,714	1,941	5.3
Laborers	17,154	1,249	7.3	9,710	936	9.6	7,444	313	4.2
Service workers	1,985	651	32.8	1,881	589	31.3	104	62	59.6
Total blue collar	92,723	5,711	6.2	47,784	3,388	7.1	44,939	2,323	5.2
Total	150,855	6,285	4.2	93,272	3,663	3.9	57,583	2,622	4.6

Source: Data in authors' possession.

TABLE A-22. *Electrical Manufacturing Industry*
Employment by Race, Sex, and Occupational Group
486 Establishments, South Region, 1966

Occupational Group	All Employees			Male			Female		
	Total	Negro	Percent Negro	Total	Negro	Percent Negro	Total	Negro	Percent Negro
Officials and managers	14,034	30	0.2	13,789	28	0.2	245	2	0.8
Professionals	24,005	148	0.6	23,581	125	0.5	424	23	5.4
Technicians	13,196	267	2.0	12,150	222	1.8	1,046	45	4.3
Sales workers	1,705	3	0.2	1,618	1	0.1	87	2	2.3
Office and clerical	24,542	499	2.0	8,286	211	2.5	16,256	288	1.8
Total white collar	77,482	947	1.2	59,424	587	1.0	18,058	360	2.0
Craftsmen	25,362	562	2.2	23,113	496	2.1	2,249	66	2.9
Operatives	101,258	8,024	7.9	45,400	4,058	8.9	55,858	3,966	7.1
Laborers	27,536	2,642	9.6	15,529	1,820	11.7	12,007	822	6.8
Service workers	3,541	1,109	31.3	3,193	989	31.0	348	120	34.5
Total blue collar	157,697	12,337	7.8	87,235	7,363	8.4	70,462	4,974	7.1
Total	235,179	13,284	5.6	146,659	7,950	5.4	88,520	5,334	6.0

Source: U.S. Equal Employment Opportunity Commission, *Job Patterns for Minorities and Women in Private Industry, 1966*, Report No. 1 (Washington: The Commission, 1968), Part II.

TABLE A-23. *Electrical Manufacturing Industry Employment by Race, Sex, and Occupational Group 624 Establishments, South Region, 1967*

Occupational Group	All Employees			Male			Female		
	Total	Negro	Percent Negro	Total	Negro	Percent Negro	Total	Negro	Percent Negro
Officials and managers	19,155	65	0.3	18,796	62	0.3	359	3	0.8
Professionals	31,396	188	0.6	30,751	169	0.5	645	19	2.9
Technicians	22,755	541	2.4	20,748	471	2.3	2,007	70	3.5
Sales workers	2,481	7	0.3	2,360	7	0.3	121	—	—
Office and clerical	32,272	809	2.5	10,761	382	3.5	21,511	427	2.0
Total white collar	108,059	1,610	1.5	83,416	1,091	1.3	24,643	519	2.1
Craftsmen	30,829	1,095	3.6	28,543	999	3.5	2,286	96	4.2
Operatives	138,576	16,863	12.2	54,759	6,676	12.2	83,817	10,187	12.2
Laborers	30,970	4,555	14.7	16,406	2,561	15.6	14,564	1,994	13.7
Service workers	3,872	1,345	34.7	3,515	1,195	34.0	357	150	42.0
Total blue collar	204,247	23,858	11.7	103,223	11,431	11.1	101,024	12,427	12.3
Total	312,306	25,468	8.2	186,639	12,522	6.7	125,667	12,946	10.3

Source: U.S. Equal Employment Opportunity Commission, *Job Patterns for Minorities and Women in Private Industry, 1967*, Report No. 2 (Washington: The Commission, 1970), Vol. I.

TABLE A-24. *Electrical Manufacturing Industry*
Employment by Race, Sex, and Occupational Group
693 Establishments, South Region, 1969

Occupational Group	All Employees			Male			Female		
	Total	Negro	Percent Negro	Total	Negro	Percent Negro	Total	Negro	Percent Negro
Officials and managers	23,441	142	0.6	23,016	134	0.6	425	8	1.9
Professionals	34,201	290	0.8	33,290	275	0.8	911	15	1.6
Technicians	23,759	768	3.2	21,236	625	2.9	2,523	143	5.7
Sales workers	2,928	15	0.5	2,493	9	0.4	435	6	1.4
Office and clerical	36,803	1,533	4.2	11,772	627	5.3	25,031	906	3.6
Total white collar	121,132	2,748	2.3	91,807	1,670	1.8	29,325	1,078	3.7
Craftsmen	36,386	1,518	4.2	33,692	1,373	4.1	2,694	145	5.4
Operatives	142,704	25,403	17.8	58,273	9,541	16.4	84,431	15,862	18.8
Laborers	34,217	6,735	19.7	16,732	3,069	18.3	17,485	3,666	21.1
Service workers	4,890	1,800	36.8	4,365	1,558	35.7	525	242	46.1
Total blue collar	218,197	35,456	16.2	113,062	15,541	13.7	105,135	19,915	18.9
Total	339,329	38,204	11.3	204,869	17,211	8.4	134,460	20,993	15.6

Source: Preliminary Equal Employment Opportunity Commission data, 1969.

Index